P9-DMI-048

life admin

# life admin

## How I Learned to Do Less, Do Better, and Live More

ELIZABETH F. EMENS

HOUGHTON MIFFLIN HARCOURT
Boston New York 2019

For information about permission to reproduce selections from this book,
write to trade.permissions@hmhco.com or to Permissions, Houghton Mifflin Harcourt Publishing Company, 3 Park Avenue, 19th Floor, New York, New York 10016.

hmhco.com

*Library of Congress Cataloging-in-Publication Data*
Names: Emens, Elizabeth F., author.
Title: Life admin : how I learned to do less, do better, and live more /
Elizabeth F. Emens.
Description: Boston : Houghton Mifflin Harcourt, 2019. |
Includes bibliographical references and index.
Identifiers: LCCN 2018024867 (print) | LCCN 2018028144 (ebook) |
ISBN 9780544558243 (ebook) | ISBN 9780544557239 (hardback) |
ISBN 9781328606709 (trade paper) | ISBN 9781328606709 (international edition) |
Subjects: LCSH: Self-actualization (Psychology) | Stress management. |
BISAC: SELF-HELP / Stress Management. | SELF-HELP / Personal Growth /
Happiness. Classification: LCC BF637.S4 (ebook) |
LCC BF637.S4 E446 2019 (print) | DDC 158.1—dc23
LC record available at https://lccn.loc.gov/2018024867

Book design by Kelly Dubeau Smydra

Printed in the United States of America
DOC 10 9 8 7 6 5 4 3 2 1

*For my children*

# Contents

*Introduction* ix

**PART I: ADMIN PROBLEMS**

1. What Is Admin? 3
2. The Costs of Admin, or Where's Your Head? 17
3. Admin Personalities, or Who Are You? 29
4. Who Does Admin?, or Is Admin for Girls? 44
5. Admin Is Sticky, or If Everybody's Doing It, How Come Some People Are Doing More of It? 55

**PART II: ADMIN SURPRISES**

6. Admin That Can Wreck You 67
7. Admin That Can Fix You 82
8. Admin Judgments 96
9. Admin Pleasures 104
10. Admin to Win Friends and Influence People 114

**PART III: ADMIN FUTURES**

11. Relationship Tips 129
12. Individual Strategies 153
13. Collective Possibilities 174

*Epilogue* 199
*Note to the Reader* 205
*Acknowledgments* 206
*Appendix A: Ideas to Try* 210
*Appendix B: Admin Personalities Quiz* 216
*Notes* 224
*Index* 251

# Introduction

This is the book I thought I didn't have time to write. It is also the book you think you don't have time to read. The reason for me, and perhaps for you, is admin.

Modern life is shaped by this unseen form of labor. Demands to do it bombard us, moment to moment, threatening to steal our focus and waste our time. This labor is often neither appreciated nor compensated. And yet no one can entirely escape it.

*What is on your admin to-do list right now?*

That is the first question I asked everyone I interviewed.[1] Now I am asking you. Before you can answer, you need to know what admin is.

Admin is the office-type work that it takes to run a life and a household. As with actual office work, this life admin involves both secretarial and managerial labor—filling out forms, scheduling doctors' appointments, sorting mail, making shopping lists, returning faulty products, paying bills and taxes, applying for government benefits or identification, making financial decisions, managing any outsourcing, and keeping track of everything that needs doing. This list covers only a fraction of the job description.

Almost any life endeavor can have an admin component. Celebrating a loved one's birthday may involve social admin—planning and hosting a party, ordering a gift, or both. Starting a workout regime, depending on your chosen exercise, might include gym admin—comparing gyms for their facilities and fees and locations, filling out forms, setting up payment. Adopting a dog inevitably means some pet admin—figuring out shots and vet appointments and alternative care when you travel. And calling your father just to say hello might well lead to co-navigating a solution to his computer problems—parent admin (or dadmin).

Everyone's admin to-do list differs. What is on your list right now?

One admin question leads to many others.

*When do you do admin?*

Demands for admin arrive by email, voicemail, and text; they pile up on the mail table; they lurk in our overflowing to-do lists. Yet there is no good time to complete these tasks. We are frequently left to confront them, as best we can, through multitasking and in stolen moments. Admin thus becomes like another job that runs alongside our work, leisure, and sleep, compromising each endeavor.

The *second shift* has become the term for women's household labor after a day's work outside the home.[2] Admin—with its pervasive presence in the margins of everything else—should be understood as everyone's *parallel shift*. Admin is like a second (or third or fourth) job we are each asked to do in the margins of our other roles.

*What would help you with admin?*

A dizzying array of books tells us how little time we have, how overwhelmed and compromised our minds have become.[3] The concept of admin I'm identifying helps us to *see* a pervasive form of invisible labor and to formulate solutions.

We can make life choices that reduce or redistribute or detoxify admin—and we can envision change at the level of law, markets, and community norms. Viewing admin as a form of labor is the first step to understanding how this work is created, how its demands find us, and how the pathways of these demands can be interrupted or redirected.

The book you are reading is the book I wish I'd read ten years ago, before I got married and had kids, before I started to set the patterns in my own family life. But even more so, this is the book I wish everyone who influences my life had read five years before that. If this book had been read by policymakers and employers, retail CEOs and creative entrepreneurs, educational administrators and, yes, several members of my family, then maybe, just maybe, I wouldn't have an admin problem to fix.

*Does admin affect your relationships in any way?*

I started writing this book to fix the admin problem in my own life. One morning after my second child was born, before I had any idea that admin would become a research project, I sat down at my computer and started making a list of all the different kinds of admin I do. I couldn't have imagined this would turn into a book; I had only just begun to realize admin was a thing at all. And, if it was a thing, it seemed to be only my thing.

I was trying to make this thing visible in my home, trying to find a way to convey to my wife just how much time I was spending on the details of our household. Beyond that, I was trying to lay the groundwork for her to identify which of these tasks she considered necessary. *Unnecessary* was the term she often applied to (the as-yet-unnamed) admin when I first started bringing up tasks in that category. The tasks only I thought necessary—like writing thank-you notes—probably weren't worth discussing further. A lost cause. But knowing which tasks she agreed were necessary would, I hope, give us a starting point for divvying those up. None of this was fun, especially as a mother of two with a more-than-full-time job that I loved.

From the start, my wife and I aspired to be full partners in parenting and housework. We outsourced as much of the latter as we could afford and as much of the former as we needed in order to do our jobs. We aimed to split the rest. Nonetheless, I felt overburdened. Slowly, I began to realize that the cause of my frustration was this invisible layer of work I was doing alongside and around everything else.

I don't know how it happened, really, how I became the principal admin Doer in our household. Perhaps it started with our wedding, which I did more to plan. We had a modest wedding—fewer than sixty people, mostly family, in a special place from my childhood: lakeside in a small Midwestern town. I thought it was so romantic that my wife already had the place in mind when she proposed. Little did I know that marrying in my special place would give me a special role in our wedding admin. (To be fair, neither did she.) My role as Doer may have started even earlier, though. She moved into *my* apartment, so I was already paying the rent and calling the handyman for repairs. The precise origin of our admin roles is unclear.

What is clear is that I have not always been a Doer. In my longest relationship before my marriage, I was definitely the Non-Doer. I was in fact so lackadaisical about admin that when my girlfriend and I moved into university housing for my new job, she called my employer's human resources people, without asking me, to make the arrangements. I was a little angry when I found out; her call somehow seemed insulting or intrusive. She pointed out that I would never have gotten around to making this call, or at least I would have stalled until it was so close to the move-in date that finding movers would have been difficult. She was right, and I ceased to be upset. But I felt newly aware of, and a little embarrassed about, my status as the Non-Doer.

And so there I sat, in the winter of 2013, a married mother of two and a reluctant Doer, and I was in a quandary. How could I convey to my wife what I was spending all those hours doing? And what could I say in our disagreement about the necessity and the value of this unseen labor? So I started listing these tasks.

The results surprised me. The list went down one page and on to the next. More and more items kept coming, and they started to sort themselves into categories. This book had begun to be written.



*Is there anyone in your life who does more admin than they should?*
I presented an early version of the ideas that became this book to colleagues a few months after the list-making began. The reaction was far more intense than I had anticipated. Law professors in the audience talked as if I had seen right into their minds and marriages. Starting with that first presentation, I have heard again and again versions of the same response: "You've given me a word to describe this thing in my life," "You've changed the vocabulary in our household," or "I've never read a paper that so perfectly describes my life." Admin seemed to be hitting a nerve.

At that first presentation I gave, an audience member told the story of a friend who once tried to list all the unseen things she did for her household—things the speaker now understood fell into my category of admin. Her friend had started trying to record all her admin endeavors in a day, and, apparently, before sundown, "she gave up."

Perhaps she felt overwhelmed by the sheer size of the list she was making. Perhaps she was afraid to admit to herself just how much more she was doing than her husband to run their household, afraid to see inequity in a relationship that aspired to equity. Perhaps this woman understood that no list, however long, would suffice. Perhaps she realized that list was really the beginning of a book.

*Do you ever talk about admin?*
Realizing that admin was a thing led me to start asking admin questions. First I asked these questions informally, of almost everyone I encountered, and then eventually through interviews and brainstorming sessions. I knew the book would be written through the lens of my life and experience, but I wanted to reach beyond

that, to learn more about the many faces of admin in our vastly different lives. I can't promise to capture your experience, but I hope to touch some aspect of it in the pages that follow.

Asking people about admin, I also started to see its lighter side. I began to get a sense of humor about how bad admin can make you feel and to find some relief in hearing other people talk about it.

*What kinds of admin would you most like not to do?*

The time we spend on admin can skyrocket during major life events—both challenging and joyful ones. Illness and impairment often come with substantial labor of this kind (disability admin), and the admin created by divorce or the death of a loved one can practically overwhelm a person already weighed down with grief (divorce admin and death admin). Job loss can trigger an avalanche of highly stressful financial admin (as well as job-hunt admin).

Happier occasions, too, can be admin-intensive. Consider wedding admin or graduation admin. The onslaught of parenting admin that accompanies a new baby rivals sleeplessness for the least fun part of the job.

*Does admin occupy space in your mind? Are there moments when admin isn't on your mind?*

Everyone above a certain age faces admin demands of one form or another. Indeed, admin may define adulthood today.

And yet school provides no training for this work, and the market, at present, offers few solutions. Vacations invite us to escape admin (even if they usually require admin to make them happen). That out-of-office autoreply may confront personal as well as professional callers. The road trip, the retreat, the allure of going off the grid—our collective escape fantasies, our nostalgia for childhood—these offer tantalizing glimpses of our lives and our minds freed from admin. But no escape route can free us entirely.

.........

Life admin almost prevented me from writing *Life Admin.* While I have been trying to write this book, I have moved my household to a different state (twice); searched for a rental home (twice, from out of state); trudged through New York City's notorious kindergarten-application process (twice, once from out of state); hired nine different babysitters (and said goodbye to most when we moved or they moved or we couldn't afford them anymore); transitioned two small children into new schools (twice for each child because of the moves); took a year-long position visiting at another university (and then returned); served jury duty (fortunately, for only two days); had my car broken into (by someone who left gloves and blood inside); and, last but certainly not least, labored through marital separation and divorce (nearly complete).

Any subset of these events could have made this a big admin era. Some of my admin, like applying to private schools in New York City for my kids, reflects my idiosyncratic demographic as well as tremendous privilege. (Let it be noted, though: While you're doing this kindergarten-application process—which required writing more essays than I wrote to get into college—it doesn't feel like nearly the privilege it is.) Other admin I've faced lately, like moving to another state for a job or renting a home, is familiar to anyone with the good fortune to have job prospects and money for rent. Other admin-intensive events in my life, like divorce, no one wishes for. Combining good and bad alike, I have faced more admin during this time than I ever imagined possible.[4]

This is not a memoir. It is not a tell-all journey through divorce admin or school applications. Or a how-to guide to solving any of the particular admin problems I've faced this year. And yet, writing this book saved my life. Had I not identified the concept of admin as a problem to be solved, had I not read innumerable books and articles on related topics, had I not spent countless hours conducting interviews and brainstorming sessions to learn how various individuals experience and address this problem, had I not found other people struggling to face their admin with energy and intelligence and collaborated with them, I surely would never have survived this admin hell, much less managed to get to yoga fairly often, lose my baby weight (and gain some back and then lose most of it again), and, eventually, stare at a blank screen long enough to write this book.

Admin's demands on my time and mental space were one major challenge of writing this book. The other major challenge was admin's demands on you. How does one write a book for people facing such impositions on their time and mental space? Your quandary has plagued me, these many months, as much as my own. I've had no choice but to make it worth our while.

*Have you developed any tools or strategies to help you with admin?* How did this project save me? The short answer is innovation. I had to invent solutions to admin. Ones I could implement in my life, as well as ones I could only fantasize about effecting in the world around me.

I gathered most of my ideas from other people. Through my interviews and brainstorming sessions, I have gained admin insights from over a hundred individuals, to whom I am deeply grateful. And there were also many more colleagues, friends, and acquaintances I didn't interview who shared their stories and ideas while I was traveling the lecture circuit, puzzling about admin during shared meals, or searching for conversation topics at my children's school events.[5]

People have divulged their secret feelings about admin. They have conveyed the texture of admin in their lives. They have shown me the notebooks and calendars and piles where they keep it. They have shared their admin aspirations, their hopes for the kind of people they wish to be in the future. They have described how they've used admin to pursue their own ambitions or help other people or avoid whatever else they should be doing. They have confessed their private manipulations around admin in their relationships, whether successful or doomed. They have narrated their attempts to change the way admin is done in their communities and their commercial encounters. They have outlined their strategies for dealing with it and living with it.

I am deeply interested in all these topics, but the last has been vital. Spurred by a researcher's fascination and a drowning person's desperation, I have become a sponge for admin strategies.

The pages that follow take us from admin problems through admin surprises to admin futures. Though the last part focuses squarely on how we and the world can change, the ideas I've gathered for how we can better address admin in our lives, individually and collectively, are offered throughout.

Understanding the problem (Part 1) means not only making admin and its costs visible and tracing the patterns by which it becomes "stuck" to people, but painting a picture of the four main admin personalities I've encountered. Illuminating some surprises (Part 2) shows admin at its worst and its best—we see here how admin can clobber us, but also how it can transform us; how admin comparisons can separate us from those we love, but also how an admin perspective can help us find new pleasures, realize goals, change our environments, and even love better. Imagining the future (Part 3) starts local by offering ideas for improving our relationships and strategies for tackling admin at an individual level. But admin is not just your problem or my problem. It is everyone's problem. The final chapter builds on this conviction and begins to fantasize about admin utopias—radical departures from our current world—before zeroing in on concrete structural changes we can make to law, markets, norms, and education.[6] The epilogue offers a snapshot of where I've landed in my own admin journey, followed by a collection of my favorite Ideas to Try. I hope this spares you the trouble—aka reader admin—of marking practical suggestions in the book as you read. Lastly, I've included an Admin Personalities Quiz, inviting you to explore your own admin ways.

*What would you be doing if you weren't doing admin?*
If you had a block of twenty extra hours this month or an extra five hours every week, what would you do with the time? Would you commit to a vigorous new endeavor—exercising, learning a language, practicing meditation or a musical instrument, joining a book club, investing in your professional development, volunteering in your community? Or would you embrace the possibility of downtime—see a friend or a movie or just get some sleep? What would you do with the time you gain from solving an admin problem?

I ask myself this question often these days. Identifying admin before it lands makes this question possible, even urgent. This is one of the biggest changes I've noticed now that admin has become visible to me: I can see it coming, flying in my direction. That gives me the chance to make a choice—and more often than I would have imagined, using strategies I discuss in the book, I can choose to save myself or someone else time and mental space. I sometimes think of those extra hours as Admin Savings Time.

In the year after I began this project, I spared a friend a big chunk of moving admin. He was taking a short-term job in New York City, and I hooked him up with a sublet. He had generously read the first draft of my legal article on admin, so he knew the nature of my thinking these days. After he chose to take the available sublet rather than going out apartment-hunting, I sent him this message: *You've just received the gift of 20+ hours of your life. What are you going to do with that time?*

In academic circles, this question can turn into a philosophical debate. "Do you really think," I've been asked, "that having less admin will lead people to do anything really meaningful? Or will people just find more admin to do?"

My answer is simple. I do imagine that, with less admin to do, some people will do something uniquely meaningful. I imagine that some other people will sleep more, or play more, or exercise more, or make love more. And, yes, some people will find more admin to do.

But maybe even the most unrelenting admin Doers will finally get around to making the family photo album they've been planning or sending a wedding gift to that cousin they felt so close to years ago instead of spending so much time doing their taxes or battling the cell-phone company. Maybe they'll do something ever so slightly closer to their own choosing.

What will you do with your Admin Savings Time? Even if reducing admin merely allows you to waste more time, I hope you do more of the kind of time wasting that makes your very particular heart sing.

# Part I

# admin problems

1

# What Is Admin?

They steal money, and I have to fill out forms. What a country this is.

—Wonder Woman in *The New Original Wonder Woman,* after she stops a robbery at gunpoint

This book is about something most people think is both trivial and boring. *What could be worse?* you may be thinking. I once shared that view of admin.

Admin is the office work of life. It is the organizing and coordinating and managing and faxing and emailing and calling and texting of our information and our lives. This is the kind of work that you can spend a whole day doing and then wonder, *Where did my day go?*

## The Biggest Tiniest Thing

Admin seems trivial. This is part of its dangerous logic. By appearing to be small and unimportant, admin rarely commands our full attention or inspires sustained protest. But anyone who is considering enrolling in an insurance plan, buying a consumer item, applying for school or financial aid, seeking a loan or government benefits, planning a wedding or a party, or moving to a new home is heading down a road lined with admin demands.

Each path will involve choices, conscious or not, whether to do admin, avoid admin, or delegate admin to someone else. For the admin we do, we face choices about how to do it, how much time to spend on it, even how to feel about it. For the admin we deflect or defer, we face choices about the timbre of our interactions around it. These are decisions about how we spend our time and about what demands we place on others' time. Few things could be more important.

Admin shapes life outcomes. Consider financial-aid forms for college. By one conservative estimate, the Free Application for Federal Student Aid (FAFSA) cost US families 30 million hours last year.[1] Help with this admin matters. A recent study showed that students who received pro bono professional assistance with the FAFSA were more likely to attend college, receive financial aid, and stay enrolled.[2] Anyone who has applied for financial aid can understand how the FAFSA might defeat some people.

Admin has financial consequences. One study estimates that approximately 20% of US households that could benefit from home-mortgage refinancing fail to do so, resulting in a forgone savings of $5.4 billion.[3] Why? Often, the study found, homeowners who received a letter offering them an appealing refinancing opportunity felt too busy to make the first phone call in response. Many failed even to open the letters. Those of us with piles of mail know how they feel.

## The Deepest Dullest Thing

Doing admin sounds like the quintessence of boring. But thinking about admin can be the opposite of boring.

Becoming aware of this thing I call admin has felt like emerging from an inspired conceptual art museum, the kind that makes you see art everywhere you look. Or like wandering into the old Tootsie Roll commercial where the cartoon boy strolls along singing, "Everything I see becomes a Tootsie Roll to me," as lampposts and trees morph into elongated chocolate candies.

I see admin everywhere. And it has become a site of growth and possibility, not just the epitome of anxiety and dullness. Admin is there for all the big events: weddings, births, birthdays, dating, vacations, holidays, new homes, funerals. It is the stuff of life and death.

"I worried that my partner might find what I'm talking about boring," Lauren (not her real name[4]) said after describing her admin to an assigned partner in a brainstorming session. "We think about admin as boring or . . . invisible, and yet," she reflected, admin "can be very high stakes. It can have high consequences. It can be really dramatic."

Admin is also intimate. It involves our calendars, our to-do lists —personal items we fear losing and often share with no one. Admin is the stuff of secrets: for some, those embarrassing piles of paper and stacks of mail; for others, the pleasure in creating order and checking off completed items.

First, we have to face it squarely. For many people, admin is a series of problems.

## Our Problems

My neighbor Steven, a distinguished and prolific writer, did not write a book last summer. He spent his summer in the probate office of a small New England town. His aging sister had died, and he became the family member in charge of her death admin. I admire his sense of duty, but I miss the book he didn't write.

Diana drove for over two years with an expired car registration. In some states, driving without a valid registration can lead to hundreds of dollars in fines, an impounded car, suspension of your license, or even jail time. How did this generally law-abiding woman come to break the law so obviously? She dreads and avoids her admin. What she didn't know until we discussed the concept of admin was how many people share her feelings.

A teacher I know, Tom, faced a financial crisis that saddled him with painful and complicated admin. Trying to support his family, Tom racked up $35,000 in credit-card debt. He has been fielding calls from debt collectors and making futile efforts to secure a bank loan. His latest dilemma is choosing among debt settlement, debt consolidation, or bankruptcy. Just learning what those terms mean took hours he could have spent working.

In 2014, the *New York Times* profiled a biochemist, Ricardo Dolmetsch, who left Stanford for Novartis to seek a biochemical treatment for autism. Dolmetsch's new research mission was inspired by his son's diagnosis. When the *Times* reporter asked how his son's autism had affected the work of his wife, Asha Nigh, a neurobiologist, Dolmetsch replied, "She can't work full time any more. *She's earned a Ph.D. in insurance*." He explained: "When you have an autistic child, you must learn how to navigate the elaborate, complicated medical system in the U.S. That requires a lot of time."[5]

These individuals have admin problems. As do I.

I have spent countless hours trying to escape admin in order to write about admin. Minor tasks this morning have included several forms of parenting admin (calling about a prescription refill for my daughter's latest kiddie plague; emailing babysitters about hours; asking someone knowledgeable what next-size-up car seat is sturdy and affordable), retail admin (ordering a three-pronged plug adaptor for my laptop; opening several UPS boxes of household supplies), social admin (proofreading my niece's cover letter for her job applications; wrapping a gift for a friend who's moving overseas), leisure admin (remembering to set the television to record a program I want to watch), and death admin (helping my mother decide how to dispose of some possessions belonging to her late husband).

Each of these tasks took only a trivial amount of time. Not so trivial was my main admin accomplishment of the morning: submitting a claim on my employee Flexible Spending Account. For that, I spent over an hour filling out the form, pulling claim reports off the insurance company web portal, updating a spreadsheet to tally this set of claims, compiling all these documents into a pdf, printing the official form to sign and then scanning the signed form and adding it to the pdf, and, finally, uploading the pdf to a website created by my employer. This took just over an hour because I know how to do it; the first time took far longer.

Why do I do this several times a year? To get some of my own money from my employer. More precisely, the government gives me the option to receive some of my salary in pretax dollars to make my spending on health care and childcare go further. The government gives me this tax savings so long as I go through this rigmarole.

Some readers are surely baffled by this behavior, especially those who live most anywhere outside the United States, or who have never had significant health-care or childcare costs, or who have enough money to not care about saving several thousand dollars in taxes.[6] But even those readers familiar with Flexible Spending Accounts, whether personally or by association, may be a little puzzled. Why is the government paying me to waste my time? Surely there's a better way.

Some might claim that this Flexible Spending Account admin is entirely sensible. The government wants to give that tax savings only to the people who need it. So, the argument goes, making people work to get the benefit helps to ration it, to funnel it to those most deserving of it.[7]

Perhaps. But are those who need the benefit the most also the organized and motivated people who have the time to take these steps? To begin to get a handle on this question, we need to be able to see admin. We need to identify it and understand it. We need to make it visible.

## What We Talk About When We Talk About Admin

For starters, we need to see admin as labor. Admin is labor in the sense that it's a means to an end—something we do not for its own sake, but in order to achieve some other goal. For example, most people would welcome the invention of a machine that could help them complete their paperwork at the doctor's office or find a retirement home for their aging parents just as effectively and in one-tenth of the time. By contrast, if you told people about a way that they could connect with their children just as well in one-tenth of the time, I think most people would be puzzled by the proposal (even if some might be intrigued). We tend to believe that the time spent with our children has a value apart from whatever outcomes it produces. Not so for admin.

More precisely, admin is a particular kind of labor. Admin is the kind of work that people do in offices. Much, though not all, admin involves less physical, less *observable* activity than most forms of traditional labor, such as growing food (in the fields), or cooking food (in the kitchen), or building things (in a factory), or selling things (in a store), or caring for people (in a home or school or hospital).

Some life admin involves doing the same things that are done in an office. Other admin is merely analogous to office work. For instance, making a grocery list or planning a child's birthday party is not precisely an activity that would be done in the workplace. But these tasks are similar to ordering inventory or planning an office event.

Admin includes both the kind of work that office managers might do and the kind of work that secretaries or administrative assistants (aka admins) might do.[8] Here are some basic types of life admin:

- Completing institutional paperwork
- Managing and coordinating schedules

- Managing inflow and outflow of mail, messages, and other communications
- Creating shopping lists and planning meals (but not cooking them)
- Maintaining the quantity and monitoring the location of supplies in the home
- Researching, completing, and following up on consumer purchases
- Managing utilities
- Handling finances
- Managing health, medical, and insurance matters
- Researching and applying for activities, schools, financial aid, and public benefits
- Keeping track of important documents
- Managing the selection, upkeep, and sale of any property, rented or owned
- Managing any outsourcing or other assistance (from plumbers to babysitters)
- Maintaining correspondence and gift exchanges
- Planning and organizing holidays, parties, and special events
- Planning and arranging transportation and travel, whether for commuting or vacation
- Keeping track of everything that needs to be done, one way or another.

These examples should give a picture of admin—the office work of life.

What admin doesn't include is *chores*. Most chores have an admin dimension, though, notably the "mental load" of that chore.[9] Making the grocery list is the admin dimension of the chore of grocery shopping; figuring out what supplies you need and when and how you will get clothes clean is the admin dimension of doing laundry. Childcare per se is not admin, but caring for children often involves multifarious forms of admin, like arranging for childcare or figuring out what food or clothes they need. (When the admin part of a chore involves meeting children's needs, then you're at the center of the diagram in figure 1, where all three forms of labor intersect.)

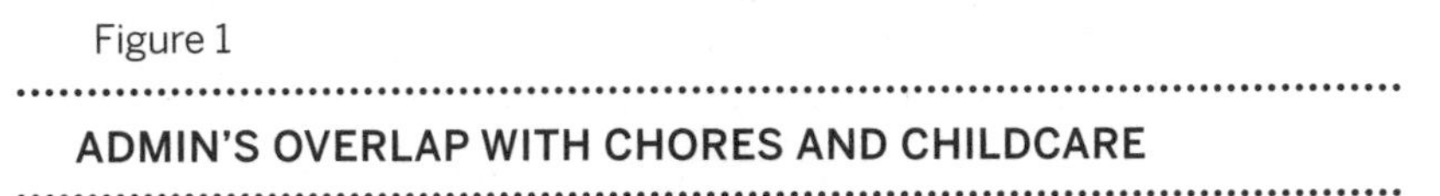
Figure 1

ADMIN'S OVERLAP WITH CHORES AND CHILDCARE

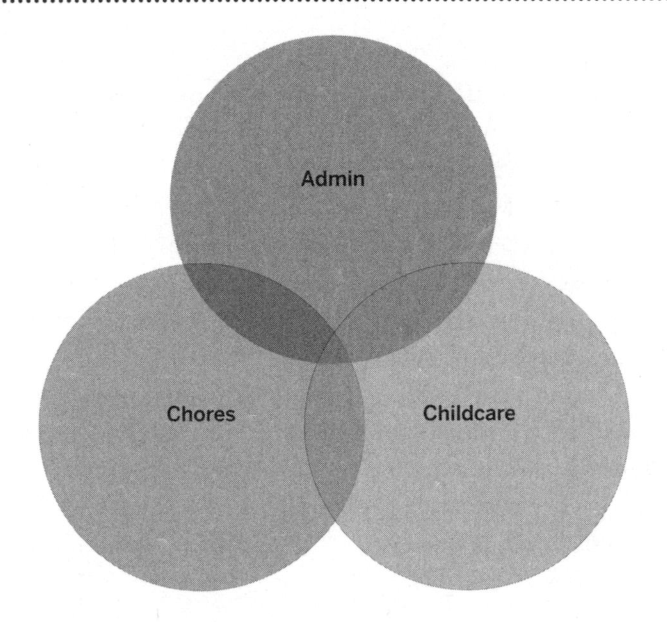


There are interesting questions about the outer boundaries of admin—what's in and what's out around the edges. For present purposes, let's not get bogged down in definitional subtleties. Most of us know admin well enough when we see it. A few people even told me they already use the word *admin* (or *administrivia* or just *stuff*) for this type of household work.[10]

At this point, I'm curious how you're feeling. After reading that list of admin examples, do you feel a little anxious or worried about some undone admin? Or does a list in that form seem remote, contained at a safe distance?

Reactions to hearing about admin vary, as I learned from brainstorming sessions. A few people found conversations about admin to be stressful, mostly because hearing about someone else's admin reminded them of more admin they needed to do. (Feel free to jot down a few things you need to do, on a list or in the margins or inside the back cover, if it helps you keep reading.) The stressed response was unusual, though. Most people found discussing admin to be a relief. It was like they were airing out some heavy blanket weighing them down. Their faces even looked lighter. Releasing admin into the atmosphere made room for other things. So let's try it.

Let's start with the worst.

## Awful Admin

If you ask people what form of admin they would use to torture someone, most will laugh at the question—at the absurdity of pairing something so serious with something that sounds so trivial—but then they may well come with suggestions.

Battling insurance companies. Doing identity-theft damage control. Looking for an important document you didn't file well (or at all). Searching for childcare. Finding someone to do something necessary, like a dentist. Facing debt or lack-of-money admin. Replacing the contents of a lost or stolen wallet. Locating lost luggage. Being audited. Filing for a restraining order. Coping with malfunctioning technology, especially phones or computers. (Like when Word suddenly started repeating a keyyyyyyyyyy and then froze while I was writing this book.) Disputing bills with the cell-phone or cable company. Applying for disability, housing, or food-stamp benefits. Taking care of anything involving the passport office.

People's ideas for the worst admin often echo their current to-do lists. What comes to mind as the worst is what comes to mind at all. Ideas for the worst admin also reflect a person's particular life circumstances. I have sometimes said, only half joking, that if you told me any person's top five forms of admin, I could locate the person demographically with remarkable precision.

High on my list of the worst, maybe because I live in New York City, is emergency-planning admin. My neighborhood is littered with posters that warn WINGING IT IS NOT AN EMERGENCY PLAN. MAKE A DISASTER PLAN WITH YOUR KIDS. After ignoring the posters for months—dismissing momentary pangs of guilt and terror as I walked past them—I tried to visit the relevant government website. The site instructed me to inform myself about *twenty-eight different kinds of disasters.* (The list begins, in alphabetical order, with Active Shooter, Bioterrorism, and Chemical Emergencies.) The eight-page Family Emergency Plan form gave me "10 Ways to Participate in America's PrepareAthon."[11] The site wanted me to learn about all these possible disasters, imagine them happening, then take specific steps for each, including filling out forms and talking about my plan "regularly" with family and friends.

My interviewees also lamented admin that's awful because it involves painful human interactions. Dealing with unreliable or dishonest or discriminatory people in consumer admin was singled out. (Related burdens include the extra admin of trying to find a trustworthy seller in notoriously discriminatory markets, or the guilt of knowing you won't find the time to do the complaining or suing-them admin.[12]) Especially awful is any admin where you are basically powerless and the result depends on other people who are indifferent to your suffering. (Government desk clerks were the prime targets here, especially when they seemed to make the rules up as they went along, otherwise known as "desk-clerk law.")

Subtler categories of awful admin emerged in my interviews. For instance, redoing admin. That is, any kind of admin that involves redoing something you've already done once—or more than once. Like when a web portal loses all the information you've just entered. (One interviewee colorfully labeled this Sisyphus admin, after the mythical king condemned to roll the same boulder up the hill over and over and over.) Or regret admin: composing apologies, paying late bills or late fees, or doing anything that is already overdue.

Participants in one brainstorming session singled out penalty-avoidance admin. Rather than offering you a possible benefit (like planning a vacation or comparing utilities for a better deal), some admin is necessary just to avoid a negative consequence (like renewing an ID card or paying a parking ticket).[13] "All you're doing is just getting yourself back to where you started and it's going to take you a whole load of time." A close cousin of these is admin cascades, where someone moves one appointment and all the events of your day, lined up like dominoes, must be rescheduled.

## Murky Admin

Murky admin is admin that lingers in your mind or on your to-do list because what needs doing is amorphous or requires you to face facts you don't want to face. You may lack crucial information. Or you're stalled on making a decision that the admin depends on.

For example, writing a letter to an old friend whose last name now escapes you. She changed it, but to what? Or calling to find out if you missed that deadline. What if you did? Do you really want to know?

Murky admin hangs out on your to-do list for days, weeks, or even months. Yet the task at hand often seems so small, tiny even, that it mocks your weakness. *You can't even write a single short email,* it says. *You can't make one little phone call.*

For a month, my identity-theft protection company called me regularly. They left voicemails. They sent me emails. They just needed an updated credit-card number to continue billing me. But I couldn't bring myself to call them back. Why? Eventually I realized I wasn't sure I wanted to continue paying for identity-theft protection. And if I did, I probably wanted to downgrade my coverage. (I definitely wanted them to stop sending me emails listing every known sex offender in my home zip code—which, in New York City, is a lot of people.) And yet it felt risky to downgrade my identity-theft protection. It took me several weeks of dodging their calls before I recognized that I had a *decision* to make—a decision I didn't really want to make—that was implicated by the simple act of calling them back. That simple act was really not so simple.

Lauren and her sister, Danielle, who both participated in a series of brainstorming sessions I conducted, immediately recognized the category of murky admin. When enough of it accumulates, they call each other and request "a consult."

Whether or not you want company on your murky-admin journey, if you want to see murky admin in action, do an inventory of your to-do list. Look for items that have been there far longer than your average turnaround time. Ask yourself if any of these are things *someone else* wants you to do—and if you don't have to do them after all. Think if any involve facts or decisions that are unknown or uncertain. Give yourself a break and treat your murky admin items as the challenge that they are. Tackle them in a strong moment, not a weak one. And bring in reinforcements. Food, drink, or love.

Surviving the most trivial form of torture deserves a celebration.

## From Suffering to Relief

That admin pain could take so many forms was a revelation from my years of admin conversations. But the biggest surprise for me was how many people seemed to enjoy admin. At least some admin, at least some of the time.

I'm not talking here about the minority who relish that archetypal admin activity: filling out forms. Those folks do exist. They have their own web groups. (Of course.) One such group is called I Love Filling Out Forms.[14] A typical post reads, "I want to work at a bank just so I can fill out forms. For the longest time I wanted to count money, fill out forms & stamp things. Is that strange?" Last I checked, that form-lovers group had just seven members, and negative comments on such pages were bold and frequent.

Many ordinary people told me, almost guiltily, that they "actually kind of enjoyed" doing their admin. Nancy named "sorting and filing" as a kind of admin torture. Then she added: "But sometimes it can be very therapeutic."

This is an intriguing puzzle: The very thing many of us avoid, or love to complain about, can indeed be therapeutic—or at least a relief. One of my hopes is that we can all find ways to convert admin from suffering to relief. When we can't get rid of it, that is.

## Is Admin the Same as Bureaucracy?

*Paperwork* and *bureaucracy* are synonymous in common parlance. "Applying for financial aid involves endless bureaucracy" can translate as "Applying for financial aid involves endless paperwork."[15] This raises the question: Is bureaucracy the cause of admin?

At its worst, a bureaucracy consists of innumerable faceless employees and a complex self-serving machinery of incomprehensible demands and detours. You, the individual, must complete many forms or make many calls that seem to lead only to more forms and more calls. Nothing is set up to make things work for you.

Bureaucracy per se is not the enemy. Bureaucracy does not necessarily create admin for people; it can lighten admin burdens through technological innovations, well-designed system improvements, and simple consideration. Here's one tiny example: The Parks Department of the county of Arlington, Virginia, collects and organizes information on summer camps for kids, making details available to parents in a way that saves them the trouble of individually repeating this exercise.[16]

What matters is whether government of whatever sort is run by people who actually think and care about the effect of their decisions on individual subjects of the state. How bureaucracy is done, rather than how much bureaucracy there is, is what matters for each of our admin lives.

2

# The Costs of Admin, or Where's Your Head?

> Life is important. Our time here is important. . . . I feel outraged that my job doesn't want me to miss like six minutes of a workday but changed my health insurance in such a way that's added at least eighty hours of annual labor on my part, on my own time.
>
> —Lauren, brainstorming-session participant

Admin seems to many people like wasting time—even killing time, as the saying goes.

If someone says, "I spent half the day at the DMV," or "I spent two hours on the phone with the insurance company," any reasonably attuned listener will groan empathetically. Active waiting—waiting on hold, for instance, or waiting for an appointment to start—is a paradigmatic example of wasted time in our culture. The admin may need to get done, but the time it takes often seems a loss of precious hours and minutes.

Admin also drains our mental energy. It can interrupt our relationships. It can even lead others to judge us—or us to judge them. And its costs vary widely among individuals.

Let's get a picture of its impact.

## Opportunity Costs

> Zing wide awake at 3:30 a.m. with thoughts like those of Anne Kimball, 46, a mother in Oxford, Pa., as she runs "down the menu, from kid to kid": "Did I send in the permission slip by deadline? Should I chaperone the field trip? Am I green enough?"
>
> —Pamela Paul, "Sleep Medication: Mother's New Little Helper"[1]

Admin consumes time that could be spent on other activities. Sleep and leisure are often first to be sacrificed. The epigraph—with the writer zinging "wide awake at 3:30 a.m."—narrates admin's imposition on sleep. Multiple people I interviewed talked about the just-before-dawn wake-up with admin on the brain. I know that one well. Admin also cuts into leisure time. Several studies have found that working women with children have significantly less leisure time than their male partners.[2] Admin is part of that gap.

Admin is often something we do when we are putting off something we consider more important. One study has coined the term *pre-crastination*—for when we do less important things too soon.[3] Admin can be a way to pre-crastinate, especially if we opt to do needless admin, such as completing consumer-satisfaction surveys or researching products that could be successfully "picked" (quickly and intuitively) rather than "chosen" (slowly and deliberatively).[4]

Admin interferes with paid work and educational pursuits. One reason illness and disability can be debilitating is that the associated admin takes time away from work and other endeavors. Cybele, a young woman with several disabilities, including cerebral palsy, described to me the laborious process of re-proving her disability to various entities to maintain her benefits, her transportation, her wheelchair, and more. She can type but she can't write, so paperwork requires her to find someone to do the writing for her, and transportation involves booking an Access-a-Ride van and then dealing with the many ways they mess up. "When I go to the doctor's office, the appointment can take like fifteen minutes, but it can take the whole day." People think impairments like cerebral palsy are limiting, but, Cybele observes, "A lot of it is bureaucracy, not really disability."[5]

In routine circumstances, the toll life admin takes may require a person to cut a few corners at work, to lose focus sometimes, or to miss out on enriching professional or educational endeavors. In extreme cases, lost work hours due to admin can hinder a person's ability to complete the tasks necessary to keep her job.

*Time is money*, yet time is not fungible. We cannot really get it back. And pragmatically, time is valued less than money, in a legal sense. Individuals generally have no recourse if companies or employers or other people burden them with time-consuming admin.[6] These damages are not measured in hours or days.

## Stealing Focus

> To do two things at once is to do neither. —Publius Syrus[7]

In addition to time, admin takes mental energy. It drains our mental resources not only when we focus squarely on it. The parallel shift can claim our attention when we are trying to do other things, like read for work or pleasure or connect with loved ones.

We all know how it feels to be in a conversation when the other person is checking her device or just mentally lost in her to-do list. Even on a phone call, we can hear that vacant quality in someone's voice—that "away," as it's been aptly called.[8] We feel distance in place of connection.

Topics that weigh on us, that cause us stress, exact a toll on our minds. Research shows that being primed to think about money when money is tight imposes a "bandwidth tax" on other mental activities.[9] Simply put, scarcity makes us less smart. Admin that relates to scarce resources may therefore be especially taxing: for instance, dealing with overdue bills or collection agencies.

The same principle may apply to being pressed for time. When time is scarce, we may be mentally compromised by a kind of tunnel vision—leading us to make more mistakes or lose perspective.[10] And our unfinished tasks occupy mental bandwidth even when we aren't thinking about them. A series of studies of the so-called Zeigarnik effect show that we remember interrupted tasks far better than tasks we've completed, suggesting that our minds continue to hold on to those undone tasks.[11] Imagine answering the phone just after putting a teakettle on the stove to boil. Even while you talk, some part of your mind remains with the heating kettle.[12]

Many of my interviewees described the feeling of minds loaded with all the admin that needs doing. As Rena put it, "You're always sort of thinking about what you need to do, what you haven't gotten done."

Multitasking is a common way to do admin. Calling the utility company while unloading the dishwasher. Filling out an online form while waiting on hold. Interviewees commented that it was hard to estimate how much time admin took because, as Rena said, "I don't usually devote time to it." It's more "catch-as-catch can." In Barbara's words, "I think you're really doing it all day long."

Multitasking prevents "pleasurable absorption into a . . . mundane task," as a psychologist I interviewed noted. Getting into a flow with any kind of work or play supports creativity and enjoyment.[13] Multitasking interrupts that. Research suggests that women in particular experience distress when multitasking at home.[14] Everyone, though, can relate to the burdens of mental distraction from competing demands. And some research suggests that no one is technically capable of multitasking (beyond truly automatic activities like walking while talking); instead, we just switch back and forth very quickly.[15]

## Why Admin Can Feel So Consuming

Major admin projects, such as buying a house, applying for subsidized housing, or planning a move, can feel overwhelming. What makes them so?

The metaphor of office work gives us a window into the problem. In any sizable office, if you did all the work involved in a particular project, then you would be assuming many different roles and responsibilities. A well-staffed office divvies up the parts of a project among multiple actors. The CEO, the secretary (or admin), and the head of human resources each have a part to play—along with many others. If you do a life-admin project alone, you are playing all those roles at once. And without pay.

## Inviting Disrepute

Concerned commentators—scholarly and popular, secular and spiritual—recommend putting away our devices to free our minds. Strategic efforts to go off the grid, whether for a dinner or a day or a whole vacation, may benefit us or those we love. *But going off the grid does not solve the problem of admin.*

If admin is pressing enough, it may still be on our minds. We may still be solving problems of planning or scheduling, or, less productively, we may be preoccupied by the prospect of forgetting what needs to be done. If admin needs doing, then putting it aside now will mean missing out on another activity later.

But doing it now may invite another kind of cost: judgment. Consider the following moment in Daniel Goleman's book *Focus*:

> The little girl's head came only up to her mother's waist as she hugged her mom and held on fiercely as they rode a ferry to a vacation island. The mother, though, didn't respond to her, or even

> seem to notice: she was absorbed in her iPad all the while. . . . The indifference of that mother . . . [is a symptom] of how technology captures our attention and disrupts our connections.[16]

Goleman surely does not intend to demonize this mother. He means to present her as symptomatic of the contemporary moment of multitasking, of distraction from human connections and what matters most.

He may be right that the mother is disconnected. Then again, maybe she is doing something that matters to her human connections. Maybe she is ordering groceries or trying to find a camp that fits her daughter's age group and interests; maybe she is researching a medical symptom her daughter just complained of or making a doctor's appointment. In other words, maybe she is doing admin.

That mother's dilemma—whether to order the groceries now and miss a precious moment with her daughter—may be part of a distributional dilemma in her household: Is this mother doing more than her share of admin? Or there may be no one else to do her family's admin: Is she the one adult in her household, carrying the weight of all the admin on her shoulders? And more broadly, this mother's dilemma points to a problem facing all of us: *When is the right time to do admin?*

When riding the subway, I sometimes read poetry on my iPhone. I told this to a friend, and now she looks at people differently on the subway. She sees them huddled over their phones and pictures them reading poetry.

These days, when I'm annoyed at someone who, transfixed by her phone, lets a door close in my face or walks directly into my path, I imagine she is doing admin. I still don't like doors in my face, and I don't really believe all the offenders are doing admin any more than my friend thinks all the subway riders are reading poetry. But the reframing decreases my impulse to judge, at least a little, which softens me. It takes the sting out.

Making admin visible has changed a family relationship for me. My stepmother is often on her phone. She has been known to text under the table at dinner. This used to bother me a lot.

Her behavior looks different to me now. My stepmom is the person in the family who finds people doctors. When anyone faces a serious ailment, she is identifying the best place to treat it and getting an appointment. Over the course of her career, she has mentored thousands of women and minority small-business owners —connecting people, giving advice, helping people put out fires —efforts for which she has won countless service awards. She pays the bills and manages almost everything in the household for herself and my father while working more than full-time. If you email her the day before Passover and ask for a recipe for brisket, one lands in your inbox in hours if not minutes. She is there when anyone needs her. Fast.

Only recently did I begin to put these two ways of being—the one annoying, the other beautiful—together. If she were tending to the sick in a hospital, her generosity would be apparent to us all. *Oh, too bad,* we would say. *She couldn't come to dinner; she's with a patient.* We'd miss her, but we would not feel bereft of her attention.

Now when I look across at her, our eyes not meeting as she taps away, I imagine a different narrative. I think of the lucky people she's tending to.

My friend Jess recently told me that the concept of admin helped her understand her (now deceased) mother's distractedness when she was a child. Jess and her siblings had often teased their mom about mixing up their names or committing other verbal faux pas. Now Jess sees how full her mother's mind must have been, managing a house, kids, and much more. I hope my own kids grant me the same understanding, someday if not now, as I mix up their names or respond too slowly or type-type-type on my own phone, communicating with their doctor or their school or Amazon and missing a moment of connection.

## The Great Divides

The costs of admin differ for different people. Two great divides separate people's experience of admin's costs.

The first divide concerns *quantity* of admin. The feeling of being burdened by admin doesn't increase in a linear way. Having a hundred admin tasks that each take one minute feels heavier than having a single admin task that takes a hundred minutes. There seems to be some threshold quantity of admin demands, and when you get above it (or trapped under it, more like), you feel burdened in a qualitatively different way. When you're in an admin onslaught—especially if it's connected to a painful life event—admin can feel so awful that the experience is unrecognizable to someone outside it.

I wrote the first draft of this book from inside a painful onslaught (divorce admin). My draft was littered with words like *torture* and *terror*. I even inserted a full-page diagram of the old Homeland Security terror-alert chart and invited the reader to locate her admin level on it. (I was often at red.)

But admin is not true torture or terror. And as I emerged from that onslaught, I revisited my draft and deleted much of the torture and the terror.

Several months later, I learned that the great divide around admin quantity is even deeper and wider than I'd imagined. I used to think that if you'd *ever* been in an admin onslaught, you could relate. But while editing the book, I went through another awful admin onslaught (more divorce admin), and it was far worse than I remembered. It turned out, even *I* couldn't remember admin hell when I wasn't in it. Not really.

The pain of admin suffering may be forgettable—much like the pain of a toothache (or even childbirth). If only we could remember these kinds of experiences keenly, when we no longer have them, we might inhabit a daily state of bliss. *No toothache,* you'd think, and feel elated.[17] *No hellish divorce admin today* would come to mind, and you'd feel like you were floating—or at least I would. But no. We are remarkably good at forgetting. The main divide around admin quantity seems to be whether or not you are *currently* in a painful admin onslaught.

The second chasm is what we might call the privilege divide. Some people's admin is awful because it's the result of poverty or lack of social capital. People of means typically do admin that involves private entities, choice, and influence over others. People without means typically do admin that involves public entities, obligations, and submission to authorities. These are very different experiences in several ways: in the tone of the interaction, the likely wait times, the stakes, and the possibility of recourse if things go awry. The costs are very different indeed.

As Esther Duflo has said:

> From our position of being reasonably well off and comfortable, [perhaps] university professors, we tend to be patronizing about the poor in a very specific sense, which is that we tend to think, "Why don't they take more responsibility for their lives?" And what we are forgetting is that the richer you are the less responsibility you need to take for your own life because everything is taken care of for you. And the poorer you are the more you have to be responsible for everything about your life. . . . If [people with money] do nothing, we are on the right track. For most of the poor, if they do nothing, they are on the wrong track.[18]

And doing the office work of life is surely that much harder if you don't have an actual office—with printer, scanner, computers, and the like—or office-work skills to draw on.

Moreover, the admin of poverty sometimes seems not just incidentally painful, but intentionally so. Humiliation admin, it might be called. Kaitlyn Greenidge writes eloquently of her memories of falling into poverty as a child, after her parents' divorce. She remembers her first trip to the grocery store, her mother presenting food stamps at their regular checkout counter, and the cashier telling them no, she couldn't help them; they had to use a different line for food stamps. "If you have ever had to deal with the bureaucracy of poverty, of having to prove over and over again to those in charge how fundamentally unworthy you are," she writes, "you understand."[19]

Seeing admin more clearly has made visible to me aspects of living in poverty that I hadn't appreciated before. One of my aims in these pages is therefore to talk across that privilege divide—to talk about vastly different kinds of admin in different lives—with the hope of illuminating an aspect of the quandary of poverty. When I started my research, I did not anticipate how hard that would be. The difficulty became apparent in a brainstorming session I conducted with advocates at a legal-services clinic. I began by asking these dedicated law students about their impoverished clients' admin, and then I turned to asking them about their own admin. These well-intentioned law students found it nearly impossible to talk about their own frustrations with scheduling the cable-TV person, for instance, after talking about their clients' struggles to keep their low-income housing or custody of their children. "It does make me feel selfish," one clinic student said, "when we're comparing our admin problems to real problems."

Despite the difficulties, I aspire to talk about admin across the quantity divide and the privilege divide. I am not saying all forms of admin suffering are equivalent. They are not. But I am also not saying that because some people's admin is surely worse than yours, whoever you are, yours shouldn't matter. I have seen, in these years of conversations around admin, the possibilities for *admin compassion*. I hope that, in risking talking about these divides and telling stories on both sides of them, maybe we can see each other a little bit better.

## The Costs of Avoidance

This chapter has focused on the costs of doing admin. But *not doing* admin can also create a variety of costs, tangible and intangible, for oneself or others.

Where particular admin is assigned to recognizable subsets of society, the costs of avoidance weigh more heavily on that group. In the family, for instance, who is called if permission slips are forgotten? Who feels responsible if bills go unpaid or investments are neglected?

The costs of not doing admin also fall differentially on various social and economic groups. My mail can pile up for a week or more. If you rely on means-tested benefits, not opening the mail for a week might mean missing a deadline that leads to losing your home.

These days, a striking example of disparate admin burdens can be seen around what are called Legal Financial Obligations (LFOs) —the small-scale citations and financial penalties that add up to big fees, repeated police stops, jail time, and worse.[20] In July 2016, Philando Castile, a thirty-two-year-old African-American man, was shot and killed by a police officer in St. Anthony, Minnesota, while his girlfriend and her four-year-old daughter sat in the car. Follow-up news stories revealed a vicious cycle of police intrusion and LFO admin hell lasting for many years prior to the shooting. Castile worked as a school cafeteria supervisor.[21] He had no felony record but had been stopped by police at least forty-six times between July 2002 and July 2016. The basic pattern was that he had traffic tickets or parking tickets he was unable to pay, leading to the suspension of his license, leading to him being stopped and charged with driving without a license, leading to more fines, and so on, ultimately leading to the encounter with the police that turned fatal.

By contrast, my interviewee Diana, a white woman, had been

driving with an expired registration for nearly two years when she was pulled over, but the police were nice to her. "It's good I'm not black," she observed when reflecting on the incident. "You hear people really have to have their papers together if they're male and black and driving, and I was able to go around with bad papers for a really long time." Eventually, five months after the police stop, Diana got around to getting a new registration.

Admin burdens—and their psychological and financial costs—vary along many dimensions of identity. They also vary depending upon your admin personality type. Whatever type you are, I hope you will get some validation, some strategies, and some relief from these pages.

3

# Admin Personalities, or Who Are You?

Pretty Obvious Which Sibling Going to Have to Deal with All the Nursing Home Stuff
—headline from *The Onion*[1]

The satirical newspaper *The Onion* ran this story several years ago. The photo portrayed a plain, competent-looking woman in her late forties standing in her suburban living room facing the camera. She stared out, sturdy, resigned. The "article" reported that her siblings all agreed that if they were being honest, "really only one of them was suited to the task" of putting their parents in a nursing home.

Her thirty-five-year-old brother summed up the family feeling: "It's a Sarah thing, for sure. She can handle those things easily enough: finding the right place, signing them up, dropping them off, stopping by regularly, making sure the bill gets paid on time. I actually think she'd kind of like doing it." The article concluded with the news that Sarah's parents were "hoping for Sarah as well, since the prospect of depending on one of their other children for care 'absolutely terrifies [them].'"

Why is this funny? For starters, the headline is about something tedious and common—the opposite of news, a trademark *On-*

*ion* maneuver. (Other admin examples: "401K Enrollment Form Sits at Bottom of Desk Drawer for 22 Years,"[2] "Man Waiting in H&R Block Lobby Nervously Eyeing How Much More Paperwork Everyone Else Brought,"[3] and, a favorite of mine, "Woman Going to Take Quick Break After Filling Out Name, Address on Tax Forms."[4])

And *The Onion* article on aging-parents admin is calling out something widely known but typically unsaid. There's often a tacit understanding in extended families about who is the Doer—and about that person's special talents or enjoyment of the work. Sarah will not just do the aging-parents admin. She will be good at it. She will even *like* it. Or so her brother tells us.

Sarah's a Doer. And her family knows it.

## What's Your Admin Personality?

Examining admin closely has led me to identify admin personalities. The basic divide is between Doers and Non-Doers. The Doers are doing it—or at least trying to. The Non-Doers are not doing it and not trying to—not really.

Parsing things a bit further, we get four types. These are, as portrayed in the four-square below, the Super Doer, the Reluctant Doer, the Admin Avoider, and the Admin Denier.

Perhaps the names of these types resonate for you. Perhaps they bring to mind people you know. Or perhaps you need more explanation. (And if you want to experiment with the types, you can take the quiz at the end of the book.)

The basic differences among the four personalities involve variations along the dimensions of action and feeling: whether a person's basic orientation is toward getting admin done or not getting it done (the *action* element) and how a person feels about the state of admin in his life (the *feeling* element).

Figure 2

**ADMIN PERSONALITIES: ACTION AND FEELING**

| | Feeling Good | Feeling Bad |
|---|---|---|
| Doing | Super Doer | Reluctant Doer |
| Not Doing | Admin Denier | Admin Avoider |





Those in the top row—Doers—are mostly getting admin done. Those in the bottom row—Non-Doers—are mostly not getting it done. Those in the left column feel pretty good about their admin, and those in the right column do not.

Many of us don't fit neatly into one box, and our personalities can change across contexts or relationships. Before delving into the nuances, though, let's get a basic picture of the characters.

## Super Doer

The Super Doer is on top of admin. She is the Wonder Woman who promptly supplies recommendations of every kind—doctors, real estate agents, hairstylists, snow removers—to her family and many friends (and even strangers), plans gatherings and trips for everyone, and remembers (or has a system for remembering) birthdays, even while successfully managing her demanding job or household of children or both.

He is the hyper-efficient Superman whose inbox is always empty by nightfall, who gets enough sleep, fits exercise into productive days filled with work or family or both, and who pays every bill the moment he opens it—or who did so until he made a calculated decision about his values, now that he has small children at home, to wait until the next day at work to pay the bill. (And lest you wonder, this guy has already calculated that the interest he loses by paying immediately is negligible these days.)

Super-doing need not mean doing all the admin yourself. It can mean being super-intentional and conscientious about outsourcing. Super Doer Erin, who juggles kids and a demanding job, had the foresight and financial means to hire a nanny to do the admin of childcare, not just the care itself: "As I told her when I hired her, *I* want to be the weak link. *You* are reminding *me* of things." The nanny does just that. The nanny remembers that one child has a birthday party to attend and needs a present. The nanny points out that the littlest of the three children needs a stool to reach the bathroom sink. "Identify a need, go to Target, buy stool, come home. I don't need to participate." Erin recognized the labor of caring-for-kids admin—the thinking and organizing and noticing and keeping track of a zillion details—and was not confused or conflicted about outsourcing that work.

The Super Doer is Nate, a politician, who styles himself "a huge fan of aggressively scheduling." Nate has been careful to surround himself with other Super Doers who bring to their joint life and work projects a "passion for being organized" and for "having the trains run on time so you can live your life." His wife, Cheryl, remarks, only half joking, "[Nate] thinks that in my former life I must have been involved in time management. . . . I'm automatically thinking five steps ahead." Ellen, whom they've employed to manage their personal and professional lives for over a decade, immediately identified herself as a Super Doer, adding, "I'm a go-to, do-it-all kind of girl." This Super Doer trio—Nate, Cheryl, and their assistant, Ellen—meet every week or so for a family scheduling meeting, where they preview the coming week and beyond, make decisions about which speaking and event invitations Nate and Cheryl should accept, and map out plans (like summer schedules) for Nate and Cheryl's kids. They schedule everything, Cheryl says, including "blocks of time to be with our family."

The Super Doer is also Rosa, who has always been a planner, unlike her parents and siblings. The first person in her family to go to college, Rosa has been translating for her immigrant parents from the age of eight, when the family moved to this country. "Planning and scheduling just comes naturally to me." Growing up, Rosa did her own admin, such as completing her FAFSA forms for college financial aid, as well as doing admin for her parents. She went to their doctors' appointments to ensure they understood their medical decisions, and when money was tight, she created budgets — budgets her parents mostly didn't follow. She continues to advise her parents on financial and other matters and looks forward to moving back to her hometown after finishing her postgraduate studies so she can "take the reins again."

One of my favorite Super Doers is a successful woman who was skeptical about the topic of this book. On first hearing what I was writing about, she remarked, "I don't really understand this book; it's like writing a book about air." I envy this woman. To her, admin is just the air of life, freely breathed in and out alongside everything else.

## Reluctant Doer

Breathing does not come so easily for the Reluctant Doer. At least, not around admin. The Reluctant Doer is more likely to experience admin as an obstacle, something he has to contend with but really doesn't want to be facing.

The Reluctant Doer feels that the to-do list is endless — and that she is always several steps behind. No major disasters befall her or her family through admin failures. The important stuff generally gets done — she is a Doer, after all — but sometimes it's late, and often it's last-minute.

Livia put it this way: "I always feel like I'm not doing enough, but I rarely feel that it's a complete disaster. I feel like I'm always sort of managing but not completely managing."

The Reluctant Doer might well consider outsourcing some of the admin. But who has the time for all the admin involved in finding help with admin, even if he can afford it? As my well-off interviewee Saul remarked about why he doesn't have a personal assistant, "If there was a button I could push and have such a person, I would do it."

For many Reluctant Doers, admin is a site of relationship friction. This is a person who really doesn't want to be doing so much admin but realizes it has to get done and fears no one else will do it. She's not at all sure how she ended up with all the pet admin; he, with all the bill paying. But who has time for the thinking and talking and negotiating—much less couples therapy—involved in trying to get someone else to share the load? Who has time for admin-redistribution admin?

One common stereotype is the marriage in which the wife is the Reluctant Doer because—she insists or he insists—the husband just isn't any good at admin. Some label this the "learned incompetence" of husbands: Is he really incompetent, or has he conveniently let things slide so she will take over? Conversely, does she really believe he is incompetent, or is she less reluctant than she admits? Another term for learned incompetence is *strategic ball-dropping*—if you drop the balls and your partner keeps picking them up, you'll learn to keep dropping them.

The Reluctant Doer who is tackling admin for which she feels unqualified may add embarrassment to her frustration and resentment. Stephanie captured this well: "It feels like everyone else kind of has their life sorted out," she said. "And I'm the one who's just like floundering. . . . That's partly why my husband and I have a lot of conflicts. Because I'm always like, 'I'm the flounderer; how can I be in charge of everything? You made me, the incompetent one, in charge.'" Even Reluctant Doers who feel competent, however, may sense that they are not uniquely competent. There's no reason why it has to be you, Rena pointed out, since anyone could do it. And so she finds herself thinking, *Why am* I *doing it?*

## Admin Avoider

The Avoider is trying to escape admin. The Avoider sees admin out there, flying right toward her, and ducks. In the words of Chris, married father of a toddler, "I think it's one of those things that—it's like a sign of being a grownup. And I try to avoid anything that makes me feel like a grownup, and therefore avoid it."

The Avoider is Diana who drove with her expired registration for nearly two and a half years. Diana leaves email unanswered and regular mail unopened. She has faced late fees and worse—even years of delay in her professional licensing—due to her admin avoidance. She generally thinks of this "stuff" as something she's bad at.

The Avoider is also the father Zack who, after enough prodding, says he'll do the grocery shopping if his wife, Livia, makes the list. Then he loses the list on three separate occasions, each time calling her at work or during a meeting to ask her to reconstruct it for him. She calls this "the disappearing-grocery-list" problem and chalks it up to a "strong unconscious reaction" to having to do something he'd never done before he met her. What makes Livia think he isn't just incompetent? When Zack gets ready for a work trip, he makes his travel arrangements and packs meticulously. In Livia's view, his selective incompetence is a perfect example of strategic ball-dropping.

The Avoider is the aging woman who still hasn't managed to sign and execute her will, three years after her adult child helped her draft it.

The Avoider is also the businessman Marcus who says that his wife, Jennifer, ends up doing most of their admin. (She agrees, in a separate interview.) Sometimes Marcus asks Jennifer to do things for him, like pick up a prescription. At other times, he says, "Simply my lack of action drives her to action; it ends up being on her plate." He describes a similar dynamic in his family of origin where his single-parent dad always wanted him to do more around the house. "I was always very resistant to doing those sorts of things" because "I wanted to be doing whatever else I was doing." On hearing the personality types, Marcus sees himself as Reluctant Doer fading in to Admin Avoider, but not an Admin Denier: "I don't deny that it exists; I just don't want to do it."

Vera, Saul's fiancée, feels "guilty, kind of like constantly guilty." She, like many Avoiders, feels shame over the consequences she's faced for her avoiding. But not all Avoiders feel very bad. Some are just mildly guilt-ridden; these folks are often simultaneously grateful for whoever *is* getting their admin done, when someone else is.

Either way, the Avoider knows admin is a problem but either doesn't want to or doesn't know how to do anything to solve it.

## Admin Denier

The Admin Denier, by contrast, doesn't think there's a problem.

The Denier is the free spirit who racks up credit-card debt that her parents periodically pay off with a shake of their heads or a slap on the wrist.

The Denier is also the successful professional who has always had someone—an assistant, a parent, a spouse—who takes care of all his admin, so he never has to see it or even really know about most of it. After I explained what admin is, John told me, "But that doesn't take any time at all." I suggested that he must be well off enough not to have to submit health-insurance claims or use a pretax-dollar health account. Oh, no; he uses those. "My assistant has a spreadsheet that she uses to do all those things. Doesn't take her any time at all." When John hired this work assistant, whom he pays extra on the side for doing his personal admin, she said in the interview, "I will take perfect care of you."

Bella thought of her husband when she heard the category Denier:

> [He] sincerely thinks that there are things that, when I raise them, he just absolutely is like, "Why are you bringing up this complicated issue that absolutely does not need to be addressed?" that I *one hundred percent* know is essential—like homeowners' insurance. We're talking, like, very big-picture things, and he's like, "Ugh, don't be silly."

One typical difference between Deniers and Avoiders is that the Avoider often says *thank you* to whoever handles the admin. The Denier may say, "Why are you wasting your time doing this thing that I don't think is important?"

Alice, who is the main admin Doer in a polyamorous household, explained the difference between her two partners: "My husband and I also live with my girlfriend, and my husband is *really* an Avoider and my girlfriend is *really* a Denier. And so he feels terrible about the stuff that he's not doing that I help with." And the girlfriend doesn't feel bad? "A little bit. But much more of the 'Eh, whatever.'" Alice's threesome means triple the admin for her.

Of the Deniers I've met, more are men. In this culture, it seems, it's hard to be a woman and not even know about admin. If a woman doesn't do her own admin, she's unlikely to be completely unaware that others are doing it for her. So while there *are* female Deniers, the privilege of obliviousness is more often reserved for men, as well as some very fortunate young people.

## A Fluid Reality

Most of us are not simply one type or another. These admin personalities are archetypes, caricatures, whereas people are complex creatures whose realities partake of diverse feelings and life strategies. Although some people squarely fit one personality, many people move from one category to another over time or across relationships. In one relationship, you might be a Doer; in another, an Avoider or even a Denier.

A remarkable woman I interviewed, Tina—a leader in her field, mother of two—made this observation about her own life: With extended family, "I never ever volunteer for recurring things that are intensive, but I'll do one-off things that I volunteer to do, where I can pay attention to it sporadically." Like figuring out how to set up automatic bill pay for her ailing father-in-law. Her husband, Seth, she realized, "does to me what I do to my other relatives." Seth likes the idea of each spouse being a "dictator" in certain areas, but he's dictator in fewer areas than she is, and "he's dictator of things that are never crisis things." Rather, "he chooses things that are kind of one-off or quarterly"—just as Tina does with her extended family, leaving the admin that makes you feel like you're being bombarded, that interrupts you with urgent matters to be decided and tasks to be done, to her siblings. (We might call this bombardment admin.)

Tina's role reversal resonated for me. I am (relatively speaking) an Avoider in my extended family. I've also shifted across time in my relationships. Although I was an Avoider in my longest premarriage relationship, I became a Reluctant Doer in my marriage. These days, as I make my way through my divorce, I am still mostly a Reluctant Doer, and yet for some admin, I aspire to Super-Doing. I search for the satisfaction, even the pleasure, that can be found in doing admin well. I am usually quick with referrals when friends are looking for doctors or other helpers, for instance.

(I hesitate to put this in print, though, lest one of my friends disagree.) And my children's education engages me, so I sometimes get organized enough to feel on top of that.

My Avoider tendencies continue to surface in some places—like with my snail mail. I dragged a white kitchen garbage bag filled with two months' worth of mail from Boston to New York when I moved. I rifled through it periodically for anything really urgent, but it took me another two months before I tackled it. I've also learned some useful strategies from the Avoiders and Deniers I've known, and I draw on those intentionally at times. Historically, admin has never felt like one of my special talents or special pleasures. In most domains, I therefore tend to feel that I'm a Doer against my will.

People also might disagree on who counts as which type. One person's Denier is another's Avoider. I have interviewed people I felt sure were Super Doers who, when I explained the types at the end of the interview, weren't prepared to call themselves Super Doers because they knew of someone who was better at this stuff. Comparisons are always possible.

You might wonder if the personality types are a hierarchy, a ladder of competence. In other words, is the Super Doer especially good at admin-doing, the Reluctant Doer slightly less good, and so on? The answer is that the personalities are defined by feeling rather than competency, but competency plays a part indirectly. To be in the top row, you have to be getting admin done. To get admin done, you need a certain level of competency. And while Reluctant Doers may be just as competent as Super Doers, their reluctance may make them less likely to invest time and energy in systems for doing admin well. And Avoiders and Deniers can also be competent at life admin, but just choosing not to do it, for a variety of reasons. So there's a relationship between competence and personality, but not a necessary one. And, as we'll learn in Chapter 12, all the categories have something to teach us.

## Learning to Say I Do for Love



I've learned a lot from people who have transformed their dominant admin personalities. Vera comes to mind first.

Vera aspires to become a Doer of admin. For most of her life she was an Admin Denier. To her, admin feels like "sticky goo . . . off-white iridescent goo." The opposite of a clean prism. A huge weight. It makes her feel claustrophobic. Panicky. Anxious. She worries that once she begins admin-doing, it could take over the rest of her day. She constantly feels guilty about admin. She feels like admin is looking at her and waiting for her to pick it up and take care of it.

When I interviewed her—on a rainy Monday afternoon one month before her wedding—this willowy thirty-something was ready to change. Now was the time, she had come to believe, to grow up and accept that no one else was going to take care of admin for her.

"What is the difference between avoiding and denying?" Vera wanted to know. An Avoider feels bad about not doing admin, I explained, whereas the Denier simply doesn't *see* this type of work in the first place. The Denier doesn't recognize admin as "a thing."[5]

"I put myself in that category," she declared. "And I'm climbing out of it."

Her fiancé, Saul, had told me a week earlier that Vera was a Denier turned Avoider. A bill for a ten-dollar copay eventually went to debt collectors, he'd said by way of explanation. Her unpaid parking tickets led to a courthouse summons.

"Recovering Denier" was the label Vera settled on for herself. She cast Saul as a Reluctant Doer. He complains, she said, but he gets it done.

Vera wants to do more than avoid. She's finding her way out of denial. Why? She knows how much better she feels after completing some pressing admin. That's one reason. She used to find paying bills scary. And physical mail was so perplexing that she'd just put it aside in a pile. Now she knows when something goes wrong, she can pick up the phone and call the relevant person. Like when her doctor's office forgot to charge her a copay and three months passed before she faced the bills they sent.

But the real reason is deeper—more romantic. Vera wants to become an admin Doer because she loves Saul. She wants to marry him. She wants to have "an adult relationship."

What is Vera doing to tackle her admin? Her main tool is a Moleskine notebook where she keeps her to-do list. She recopies the list periodically. It seems inefficient, all that copying and recopying, but she has become much more organized, she says, since she started this to-do-list format. She feels safer with it on paper than in the computer, where the file might get lost. Having the list on paper gets the admin out of her head more fully.

Her journey is incomplete. The Future Vera will be even more organized, she predicts. The Future Vera will have really effective to-do-list apps and shared calendar programs, for starters. Seems to her the market hasn't developed adequate versions yet.

Does she aspire to become the ideal housewife, an invisible angel seamlessly doing all of her own and her husband's admin? Apparently not. She enjoys her career as a stylist. She talks pointedly about the things she and Saul are each best at in the household. She speaks in terms of "stamina." Hers is for organizing physical items and spaces. His is for admin, which permits him to handle the doctors and other health and diet details for their sick pet. With that, she gets overloaded really easily. These are moments when she really appreciates Saul.

That appreciation is perhaps the biggest change to date. She can articulate how it feels when he does admin for her. When he updates her phone, it feels "so great," not least because he does it so much better than she would. But also, she's learned over time, service is his language of love.[6] When he does her admin, that is his way of showing he cares. Early in their relationship, she wouldn't even notice him doing those things. Now she translates his acts of admin service into words of love, so she can feel the love he is giving her. She told him her translation once—told him the words she feels he is saying when he helps with her admin—and he told her, *Yes! That's exactly what I mean.*

## Changing Admin Personalities After Love

Challenging life events can also bring admin personality shifts. Pat taught me this.

After her husband died, Pat said, "It became clear there was a lot of stuff he did that I would have to do." She faced a crossroads. She felt the crushing blow of everything landing on her shoulders. She knew admin did not come naturally to her.

Pat chose to become a Super Doer.

She took steps to simplify her life, like moving from a house to a full-service apartment, where others handle everything from snow removal to lightbulb-changing. She also got help with key admin that remains: She interviewed several financial advisers and chose one. She was fortunate to have the money to make these choices.

Most striking, though, is the system she developed for tracking her time. Every morning, Pat writes her plan for the day. And then, hour by hour, in a special notebook, she keeps track of what she does in six categories, to be sure that she spends enough time on the important categories like Writing (where she makes most of her money) and that she contains her time spent in other areas like Politics (which could consume her if she let it). She gives herself extra time for doing what she calls House items, apparently to motivate herself—a full hour just for getting the mail each day, sorting it, and handling any bills right away. The tallying time she doesn't even think to count; it's just something she needs to do—she even said she enjoys it. This system is working for her: She is getting her admin done and feeling good about it.

A big life event can also inspire a personality shift in the other direction.

Lillian started to avoid admin when her marriage ended. Before her divorce, Lillian told me, she did 90% of the family's admin; by the time I interviewed her, she was doing 40% of the admin she and her ex still shared because of the kids. Her shift was partly due to how the admin with her ex feels: "Icky and emotionally triggering and sad." So she began to avoid it.

But Lillian's shift to becoming an Avoider was also driven by a conscious reflection on her values. She decided that when she is with her kids, now only half the time, she does not want mental distractions. She's come to believe that whether her children thrive depends far more on her "being a really present, loving parent" than on "whether they got into the right soccer camp." She therefore leaves her work to-do list at work, on a paper notepad, and she often leaves her phone in her bag. Her ex-husband, by contrast, now seems eager to do the admin, perhaps to maintain their connection or to prove he's good at it.

Like Lillian's ex, a divorced man I interviewed named Kurt became more of a Doer after his marriage ended. I have seen no data on this, but I suspect that when people with kids split, this shift is not uncommon. One person starts doing more, either by necessity, because he or she now has to manage a household alone, or by design, because it serves the person's interest, consciously or not, in the interpersonal dynamics of what one divorce lawyer has called "aftermarriage."[7]

Of the stories I've heard and read about one partner becoming more of a Doer when a man and a woman divorce, that partner is usually the man. This brings us to one of the biggest admin quandaries: the gender of admin.

# 4

# Who Does Admin?, or Is Admin for Girls?

This thought of independent living kind of scares me—simple things, like I'm not really sure how toothpaste shows up. Do you have to get a prescription? Is this an over-the-counter product? It just shows up in my drawer. I don't know anything about this.

—Ken Herman, a local journalist preparing to cover national politics[1]

Most of all, she keeps the literal and mental lists. . . . The keeping of those lists, [both partners] agree, makes her the de facto C.E.O. of the . . . family. —Lisa Belkin, "When Mom and Dad Share It All"[2]

If admin were a movie, would women play all the lead roles?

Before you say, "Well, of course," consider some surprising numbers. The American Time Use Survey (ATUS) asked men and women to describe how they spent their time on the day before they were interviewed. Men claimed to spend 7.2 minutes per day on "household management," a subset of admin; women, 9.6 minutes.[3] Can that be true?

Another answer: Standing at a welcome dinner for preschool parents, I mention the topic of the book I'm writing. The man nearest me says, "You should talk to my wife." This becomes a pattern—many people seem to assume that the topic of life admin is of interest mainly, or only, to women.

Still another answer comes from a male academic and father of young children who proclaimed, when I was giving a talk at his university: "Admin is not just a burden on marriage; admin *is* marriage."

Can the ATUS numbers be accurate? Can men and women both spend such a similar amount of time (and so little time) on this form of admin? The ATUS has several limitations, most notably that it relies on people's memories of how they spent their time and it largely fails to count activities done simultaneously—rather significant problems for measuring the parallel shift.[4] So what can we learn from other studies?

Researchers have observed that household management—the "essential planning, coordinating, and budgeting . . . above and beyond the physical demands of household work"—is "the last barrier to gender-egalitarian marriages."[5] In other words, as feminists have observed for decades, "some tasks are easier to delegate than others,"[6] and management tasks are particularly hard to delegate (though not impossible). Researchers have also said that household management is "the least researched aspect in the allocation of household labor."[7] So, as with the ATUS, much of the quantitative work in this area is less helpful than we might hope.[8] That said, the work thus far helps shed some light on the gender distribution of admin.

A 2015 Pew Foundation survey, for instance, found that in two-parent households where both parents work full-time, a majority of women and men say they "share equally" in "household chores and responsibilities" and "playing or doing activities with the children" (chores and childcare), but a majority say that the "mother does more" of "managing the children's schedule and activities" (admin).[9] The Pew survey found, not unusually in studies of household labor, that men and women disagree on how much household labor they each do, with women more often reporting they do more, and men more often reporting that parents share equally.[10] Even with this gap, the labor of "managing the children's schedule and activities"—a subset of what we might call kid admin or just kidmin—is the one category that both mothers (64%) and fathers (53%) agreed that mothers do more of.[11]

The spouses the *New York Times* chose to profile when reporting on the Pew survey "try to divide child care equally."[12] He "handles the morning routine"; she presumably handles other childcare time. She also calls herself "the advance team" for doing things like "ordering baby supplies, cooking meals for the week on Sundays and booking pediatrician appointments and swim lessons." Ordering and booking things are admin and, while cooking meals is a chore, planning out the week's meals is a major form of household admin.

"The Amazon Prime account is mine," she says. "He was like, 'When are we going to run out of these night diapers?' and I was like, 'We've already reordered those six times.'" She does the admin task of ordering the supplies and carries the mental load of keeping track. This dynamic between mothers and fathers seems to be common, with her doing more kidmin and meal planning, even when they both work outside the home.[13]

## Gifts and Money

Some forms of admin are especially likely to be characterized as women's work. Social admin is a prime example: party planning and inviting and RSVPing; remembering birthdays and anniversaries; keeping up extended family relationships through letters, calls, gifts.[14]

By contrast, handling finances might sound like a traditionally masculine domain. But the research suggests a more complicated picture. These days women more often pay the bills, as part of their stereotypical role of feeding and clothing the family.[15] As to who makes financial decisions, the limited data reveal no clear trend toward either men or women.[16] More specifically, there is reason to think that men more often handle money when there is money to burn (or invest), but women take over when there is a need to make ends meet; for instance, at the point of bankruptcy.[17]

Of the people I talked to, men married to women were more likely to pay bills or handle the taxes than do childcare admin. Nonetheless, financial admin was often done by women in part or in full. For Marcus and Jennifer, for instance, both partners initially credited him with doing their finances, even though it emerged that she pays nearly all their bills, including their rent. For Rebecca and Matthew, she does the financial admin, by her account, because she is "better with technology" than her husband; he is perfectly capable of using a computer, just less agile than she.

Several couples had stories about trying to share financial admin early in their relationship—for instance, trading off bill paying month to month—until one partner, usually a man, let things lapse on his turn. Then consequences would come. The internet was turned off, which they both relied on to work from home. Or a notice came that the electricity was going to be turned off. Or that the car insurance was going to be canceled. And he said some version of, *I just didn't get around to it.* And so she took over. "I knew it was a trap," Dorothy said, "but I felt like I can't live with that kind of uncertainty." Sounds like strategic ball-dropping, at least to Dorothy.

Erin bargained up front for Paul's role in their financial admin: "My deal with my husband when we got married is that I said I would not get married unless he agreed to do the taxes. He said yes. So, fine, whenever he complains about the taxes, I remind him of that earlier agreement." Would he have accepted a similar proposal for the admin of managing playdates? Perhaps not. Paul has opted out of playdate admin, Erin says. It's all mothers organizing the "playdate setup" and "they don't want to hear from the kids' dad."



## The Outsourcing Gender Gap

Outsourcing tends to exacerbate the admin-doing gender gap. Generally speaking, when a male-female couple decides to outsource, the woman ends up doing an even greater share of their household labor than before. How does that happen? For both "masculine" and "feminine" household jobs, men married to women are less likely to manage the outsourcing than to do the underlying jobs.

This is the *outsourcing gender gap.* The point is complex, so we'll take it step by step.

For starters, something interesting happens when stereotypically masculine jobs, such as home repairs or car repairs, get outsourced. Men are more likely to do these jobs directly and, if the jobs get outsourced, men are more likely than women to be in charge of outsourcing them.[18] But once these jobs are outsourced, women are more likely to be involved (in the admin of managing the outsourcing) than they were before (in the actual repairs).[19]

By contrast, for stereotypically feminine jobs, like cleaning and childcare, outsourcing doesn't shift the work away from women. Nothing suggests these become any less women's work when the task is finding someone else to do them. Rather, the woman is typically doing the outsourcing admin for the work she would have been doing herself. And in fact, for "feminine" tasks like childcare, she may be even more likely to do the outsourcing admin than she was to do the underlying work.[20]

So for example, women married to men are more likely to pick up the phone (to call a repair person) than to pick up the hammer (even though men are more likely than women to do both). And women are *even more* likely than their husbands to pick up the phone (to call a babysitter) than to pick up the child (even though women are more likely to do both). So for both kinds of jobs, "masculine" and "feminine," outsourcing nudges work more to her plate than to his.

A hypothetical case may help illustrate just how it happens that outsourcing can shift an existing imbalance further toward burdening women with a greater percentage of the household labor.

Imagine a couple, Sue and Sid, who have two kids and both work outside the home. Sue is doing 75% of the current household labor (including chores, childcare, and admin) and Sid is doing 25%. So if their total hours spent on household labor came to 60 hours per week, that'd be 45 hours for Sue and 15 hours for Sid. Now imagine that they decide to outsource some childcare and house cleaning to cover 40 hours of their previous 60 hours. And let's say the admin of managing the outsourcing takes 5 hours a week, on average, including start-up costs (like search and hiring), ongoing costs (like scheduling, communication, and payroll), and back-end costs (like termination and references for other jobs).

At this point, Sue and Sid have 25 total hours of household labor to share between them (20 hours remaining of what they had before hiring help, plus 5 new hours of outsourcing admin). If Sue does all 5 hours of managing the outsourcing, plus 75% of the remaining 20 hours, she will be doing much less overall household labor than before the outsourcing started (20 hours instead of 45 hours), but now she will be doing *an even bigger share* (80%) compared to Sid (who now does 5 total hours, or 20%). Note also that if they outsource any of their stereotypically masculine housework, like car or home repairs, then even more work may drift from Sid to Sue.

Admin's invisibility compounds the problem. Since outsourcing admin is often invisible, Sid may see only that they both have less work to do around the house. He may therefore find it hard to understand if Sue isn't happier about how much they're now doing.

What does this mean? For starters, if you're striving for some kind of balance in who does what at home, be cautious in outsourcing, even if you can afford it. If you and your partner are outsourcing your lives, and you don't make careful choices about who does the admin of managing that outsourcing, you may end up with even more of a gender skew in the work that remains than if you did everything yourselves.

## Homo Utopia?

Do same-sex couples avoid admin inequities because they inhabit a gender-free utopia? Research on who does admin in same-sex couples is even more limited than for different-sex couples, but adding anecdote to what research there is, I think we can fairly conclude that the answer is no.

Same-sex couples do appear to split the work of managing their households more equally, consistent with their general trend toward more equal sharing of household chores.[21] Idiosyncratic splits undoubtedly develop for particular couples, but a few patterns emerge.[22]

One trend is that the same-sex partner with the higher income is more likely to be responsible for financial matters.[23] Another trend applies only to women with kids: In couples where one partner carries the baby, she tends to do more childcare and, likely by extension, more childcare management.[24]

Moreover, gender if not sex may play a role in how admin is distributed between some same-sex partners. Gender theorists of the 1970s drew a distinction—conversationally useful, at least—between sex (meaning *biology*, as in male and female) and gender (meaning *culture*, as in masculine and feminine).[25] For some same-sex couples, *gendered* roles may influence admin distribution, even in the absence of *sex* distinctions between the partners. And even for the many same-sex couples who don't take gendered roles, outsiders may nonetheless expect such roles and even impose them through admin.

As one example, same-sex couples with kids will sometimes remark that for admin purposes, other parents try to figure out who in their couple is the "mom." The other parents may have a practical reason for this prying: They want to know whom to contact about playdates, parties, and parent participation in school activities. But why not just contact both parents and see who responds? Talking about admin is, for many people, an embarrassment. Many of my interviewees felt a need to apologize for talking about admin—*even though I was interviewing them about admin.* Emailing a playdate invitation to someone who considers admin beneath his pay grade feels like a faux pas, an insult to the addressee that simultaneously reveals the social isolation of the sender. It's like addressing the boss as if she's the secretary.

Last, it's worth noting that same-sex couples' reputation for more equal sharing comes with its own challenges. Couples of all stripes can fall prey to a gap between rhetoric and reality, but one researcher suggests that the dynamic of tending to "assert equality" despite "unequal divisions of household labor . . . may be even stronger" among same-sex couples.[26]

## How We Talk About the Gender of Admin When We Do

How do we usually understand the gender divide around admin?

Is it that women are just "better at this stuff"?[27] Or that men learn to be incompetent?

The idea of female superiority in admin ties in with the "women as multitaskers" thesis. Many people believe that women are better at multitasking, and the popular press often echoes this belief.[28] The empirical work on the subject almost invariably reaches a different conclusion: women are not inherently better at it.[29] Women do appear to *do* more multitasking, however, and perhaps even to suffer more through it.[30]

We've already encountered the alternative idea that men *learn* to be incompetent, that they engage in strategic ball-dropping. Some say the learned incompetence of men results from female dominance: "She just won't let me do any of this stuff." Formally, this is called *maternal gatekeeping*, the psychological term for "a reluctance to relinquish family responsibility by setting rigid standards . . . a desire to validate a maternal identity, and . . . differentiated conceptions of family roles."[31] As one writer puts it, "Many women will also admit to the frisson of superiority, of a particular form of gratification, when they are the more competent parent."[32]

Surely some women do get satisfaction from managing the household. But the female-dominance frame tends to assume that, if there is power in being in charge of admin, then women who want that power need to keep doing all that admin. Problem solved.

Understanding admin as akin to office work illuminates the quandary that remains. At the office, the decision-maker doesn't need to do all the grunt work in order to exercise power. The decision-maker need not be the person who identifies options, researches them, records and organizes the information gathered, schedules the relevant conversations and events, and then figures out how to implement any decision. Rather, the decision-maker can delegate some or all of those tasks. A committee or another individual can do the arranging and researching and implementing, leaving the decision-maker to focus on the big decisions. I'm not proposing that couples split things up in this way, with one decision-maker and one assistant. But the office work analogy helps show that being the decision-maker does not *require* also being the assistant.

In the epigraph to this chapter, Lisa Belkin tells us about the couple who agrees that the wife's keeping all the household lists makes her the CEO of the family. *Perhaps.* To some ears, though, she might sound more like the secretary.

Powerful people sometimes say that their secretaries at work are also very powerful. Their secretaries have so much information about them, their schedules, and their business that if these assistants went over to the Dark Side, they could really mess up their bosses' lives. That may be true, but does that mean powerful people want to *be* their secretaries? There's more than one kind of power, and most people with secretaries don't want the kind of power their secretaries have. To say that doing admin involves exercising some power doesn't tell us whether that kind of power is desirable.[33]

My interviewee Erin nicely captured this subtle point when I asked if there were any forms of admin she wouldn't outsource even if she had the ideal robot or personal assistant. "Anything where I feel like I'm making a judgment," she said, "about something that I consider important, I would like to be involved. But often I feel like the judgment part is relatively small, and the actual implementation is much larger."

## Should We Care?

Those who aspire to equity between the sexes may find it unfair if women are more burdened by life admin, which is often deemed trivial and taxing.[34]

The invisibility of admin also raises the concern that couples won't discuss it and so women (or other Doers) may not be clearly consenting to their disproportionate burden.[35] Note that admin is invisible not just because multitasking, especially on a device, is literally harder to see. Admin is also not widely viewed as labor. In television and movies — where we see disagreements about who cares for kids or does laundry — we rarely see disagreements about who schedules the pediatrician visit or who communicates with the landlord. I'm not saying this would be good television — though those arguments can get pretty heated! — but it's worth noting that admin's relative absence from the media makes us less likely to see it as a thing. (Notable exceptions include the British sci-fi show *Black Mirror*, which imagines a world in which you can clone your consciousness to create a perfect smart-home personal assistant, and *Seinfeld* and *Curb Your Enthusiasm*, which draw humor from accentuating trivial events like returning a library book or dealing with the TiVo repair person.)

Admin's invisibility may also lead to misunderstanding or resentment. Let's say a couple, Ben and Bert, explicitly agree to a certain division of labor in their household in terms of chores and childcare. It might be 50/50; it might be 75/25; it might be 90/10. If they have no conversation about admin distribution, but admin mostly ends up being done by one partner—here, let's say it falls to Ben—then they may have a problem. Ben spends additional hours every week on admin for the family, but because the work is relatively invisible, Bert does not realize that Ben is doing so much additional work. Ben may feel angry or resentful, and Bert may be confused. Bert holds up his end of the bargain—he does his 50% or 25% or 10% as agreed—so he does not understand why Ben still treats him like someone who is not carrying his weight.

It is easy to see why, if the bulk of household admin inadvertently lands on one partner's shoulders, the Doer may feel put-upon. Note, however, that even if the work falls equally on both, admin's invisibility may mean that neither recognizes how much the other is doing, and, thus, both may feel unappreciated.

The more I delved into people's marriages and households, the more it became clear that patterns can change, but most often, they arose early on, without a lot of discussion, and they deepened over time. Social norms and gender roles may help to explain how the patterns start, but why do they so often persist, for couples and others, even among those who value equal sharing of household labor? This brings us to stickiness.

5

# Admin Is Sticky, or If Everybody's Doing It, How Come Some People Are Doing More of It?

[Admin] feels like goo, sticky goo . . . off-white iridescent goo that just seeps into everything. —Vera, a month before her wedding

Couples aren't the only ones who divvy up admin.

Recall my neighbor Steven, the author who handled the probate admin after the death of his sister. Think about who arranges reunions or holiday gatherings in your extended family. Or who among your friends plans trips and makes reservations.

Parents generally do the admin for their children, but for how long? If the millennial generation is extending childhood longer than ever before, then in a digital era, that means parents can keep doing their adult children's admin. *Adulting* is a term popular among twenty- and thirty-somethings—as in, "I've been adulting really hard this morning." Adulting involves a lot of admin. But well into adult life, children may rely on their parents' paying their taxes, managing cell-phone or Netflix "family" subscriptions, or handling their health-insurance paperwork.[1]

The millennials' relative freedom from admin was driven home to me when presenting this project to an audience of law students.

On perusing a handout listing types of admin, one student observed, "This looks like a lot of stuff my parents do."

How do these distributions happen? Some of it is gender; some of it is personality. But neither of those explains it all. Another part of the story involves a feature of admin itself. Admin tends to "stick" where it lands.

## Being Stuck

Consider Lauren. Lauren is the principal admin Doer in the apartment she shares with several twenty- and thirty-somethings. She's been trying to spread the work around.

Her housemate Sam is always saying he "doesn't have anything to do." (Lauren doesn't have this problem, she notes with a laugh.) So when some birds pecked a hole in their bathroom ceiling, Lauren asked Sam to take charge of Project Pigeon. The Project had two parts: Cover the hole and contact the landlord to request a permanent fix.

The hole-covering part Sam had finished by the time Lauren got home from work. But the landlord part "boomeranged" back to her, as she aptly put it. Sam had reached out to the landlord, who then contacted Lauren. Lauren wasn't surprised. She had lived there the longest and the landlord saw her as the point person.

What's striking here is how difficult it was for Lauren to hand off the admin. The manual-labor part of the job she could delegate readily, once she had a willing worker. But the admin part of the job stuck to her.

## Sticky Defaults

Some would call this feature of admin *path dependence* or *status quo bias*. Or just plain inertia.

The term *stickiness* has come into vogue in behavioral economics. The idea is that when people face choices, they often choose the default rule, the background rule; they go with whatever happens if they don't make an active choice. In other words, the default is "sticky."

On matters as serious as retirement contributions and organ donation, research shows the default rule matters for people's choices. Making retirement contributions or organ donation an "opt-out" decision rather than an "opt-in" one has a sizable effect on the number of people who contribute.[2] What makes default rules sticky? One theory is that individuals attribute expertise or knowledge to the rule setter, so they infer the rightness of the choice from the way the default is set. Another theory is that people are reluctant to depart from the status quo because they feel losses more acutely than gains. A third is that people want to avoid the effort of making a decision.[3]

A related possibility is that even the admin of filling out a form is sometimes too taxing. This explanation works for a subset of sticky default findings, but not all. In some cases, the person is merely checking a box on a form she is already filling out, so there is not really any more admin to opting out of the default than to staying in. Whatever the reason, default options are sticky.

## Admin's Special Stickiness

Is stickiness a feature of all work, or is there something special about admin?

The person who starts out in a relationship cooking or doing the dishes may remain in that job in perpetuity. The same thing is true, if not more so, for taking care of children. People who favor equalizing parental leave observe that those early weeks and months in a child's life set patterns for parenting that often continue. Path dependence is a feature of all household labor, it seems.

Yet admin is particularly sticky. And the admin component of a chore or childcare tends to be particularly sticky. Consider grocery shopping, a traditional chore with an admin overlay. Taking turns going to the grocery store, list in hand, is not so difficult. But shifting responsibility for making the grocery list—the admin part of the job—is harder.

Several features of admin contribute to its special stickiness.

For starters, admin's *invisibility* makes it easier for inequities to arise and go unnoticed or undisputed and thus to persist. Relevant here is the way admin is often literally harder to see than more physical forms of labor, and also, the fact that it is not widely viewed as labor.

Admin is also *flexible*. In our technological age, much admin can be done almost anywhere. When the person who does dishes is away from home, someone else does the dishes or they pile up in the sink. But when the person who pays the bills or orders household supplies is away, he can often keep doing those tasks from afar. As one colleague's husband said before her work trip to Paris, "You can keep doing all this stuff from your phone, right?" Admin's flexibility means that even temporary separations don't force experiments in redistribution.

Moreover, admin often requires specific *information*, making it harder to take turns than with cooking, cleaning, or childcare. Someone who is a competent cleaner can clean in weeks two and four while someone else cleans in weeks one and three. (Competence—*skill* at doing the thing—matters both to traditional chores and to admin.) By contrast, on the admin side, someone who is competent at arranging playdates cannot, in principle, just jump in to arrange the playdates in weeks two and four. The person needs up-to-date information about past events and about whom to contact (and how). Skill isn't enough for doing most admin; you need to *know* things and, often, you need to know people. So we frequently hear from an overworked householder, *It's easier just to do it myself.* That may feel true for anything we're in the habit of, but it's likely to be especially true for a task involving information transfer.

Finally, people don't generally value admin in its own right. This *trivializing* of admin can make it difficult to share. Between parents, for instance, equal time with the children can be a win-win for everyone, since both parents and kids arguably get a better relationship out of it. It is harder to sell parents on the idea that they somehow benefit from doing more of the household admin. Thus, the allocation of admin work within relationships is a more bare-bones fairness issue, since spreading it around has no obvious upside for the one who has to accept more of the load.

For these reasons, whoever starts doing the admin in a relationship often gets stuck with it. This is true not only of many female partners of men, but also of one adult member of extended families and friendship networks (who organizes celebrations?), some parents of adult children and children of aging parents (who takes those distress calls at all hours?), one partner in many same-sex couples (who pays the bills and schedules date nights?), and, yes, some men partnered with women.

The term *sticky* also seems to capture something about the texture of admin. Recall that Vera described admin as "sticky goo . . . off-white iridescent goo." Not everyone thinks it shimmers. But many of my interviewees, like Vera, saw admin as something that just won't leave them alone. According to Chris, a new father, admin is like "a mosquito in the room."

## Where Admin Lands

If admin is sticky, then its starting point is important. What launches the initial distribution of admin?

1. ***Accidents happen.*** Admin's first landing may be random, falling on, for instance, whoever first bumps into the neigh-

bor with the noise complaint in the hallway. Or it may be an artifact of how tasks were split up at some foundational moment. During a cross-country move a decade ago, Rena and her husband, Kevin, took turns driving and calling to set up utilities. "There were times when I was driving and he was on the phone calling the company and he used his name; and then he was driving and then I was on the phone calling some company." To this day, Rena and Kevin still split up who pays which utility bills according to who set up which ones on that drive.

Sometimes the accident leading to admin doing is a literal injury. That's how John got stuck with the pill sorting in his marriage—when his wife had foot surgery several years ago, she couldn't walk to the pill shelf; ever since, if John tries to push pill sorting back her way, she scrunches up her face cutely and says, "I don't do that."

2. ***Punishing the volunteer.*** In other cases, admin's landing may seem to be a poor reward for the willingness to take an initial step on a household project—like calling an exterminator or installing an app to order take-out for dinner one night.

And sometimes, admin's first landing may be more predictable or conventional.

3. ***Outsiders make assumptions.*** Other people may rely on gender stereotypes to impose admin on one family member or friend. Parents see this often. Rena lamented, "I just got an RSVP to a birthday party that was only sent to the mothers." My friend Mindy reports that her husband gets contacted about giving money to their children's school, whereas she gets contacted about volunteering her time. The outside world is giving Rena and Mindy no choice in the first landing point of admin. (And if they want to try to unstick it, they face what we might call awkward admin; telling the host

what's wrong with her invite seems a poor way to get your child invited back.)

The stickiness of admin has some other predictable pathways.

4. ***Pre-relationship history matters.*** If your partner moves into your apartment, home admin is likely to stick to you. Consider Richard and Alan. Alan handles everything about renting their apartment on Airbnb when they go away. Why? There are several possible explanations. For Richard, it's because Alan had done that before Richard moved in—once. "But even having done it once, he had it all set up, and so," Richard noted, with a little laugh, "I don't think he minds doing it, and I have no interest in learning." Once was enough for Airbnb admin to stick to Alan.

5. ***Information can make you vulnerable.*** The family member who keeps records has the necessary information to make the first call. With a hint of pride, my interviewee Rebecca noted that her mother "was known by everyone to be that person."

   Rebecca's story also offers a way out of this pathway: If you keep clear enough records, other people can use them—and un-stick you. Her mother had "a fifteen-page address list, which we still have, we still use, for the house," although her mother is now deceased. "You could look under *H* for House, [and] under *E* for Electrician. Under House, there would be Electrician, and under Electrician, there would be the guy's name"—with a note: "'But make sure you ask for so-and-so; he did a good job when the front porch light went out.'"

6. ***Skills are a runway too.*** Admin's first landing may also follow skills. The person who knows how to use a scanner will likely end up doing the scanning. The skills don't have to be

great or expert, just perceived as better than the other person's.

The person who knows how to speak the language where a family lives or travels will likely face more admin landings. This is not always the case, of course. When admin-doing lands or stays with the partner less skilled in the language or techniques needed for the task (aka out-of-my-league admin), frustrations may rise. Expat interviewee Livia talked about this. She does more admin when she and her husband, Zack, visit her family in Europe, and she also winds up with more admin when they live their regular lives in America as well.

As Livia's example shows, the typical landing pathways *need not* determine who does admin. But these entry points —accidents, volunteerism, gender, history, information, and skills—often start admin down a road where it's likely to stick. Unless, of course, someone or something disrupts it.

## Stickiness Cuts Both Ways

I may be giving you ideas for how to avoid admin. *Don't answer the phone! Don't open the mail! Don't talk to the neighbors!* If you avoid all those things, you may avoid a substantial amount of admin.

Same goes for your skill set. Never learn to use the scanner! Never learn to type! This last injunction is actually the name of a book and was popular advice for professional women before the rise of personal computers.[4] If they aspired to be more than secretaries, they shouldn't learn the skills of secretaries.

As long as someone else will do your admin if you don't, you might not face any major consequences.

Or you might. Not opening the mail may be risky if you live alone or if no one else will pick up the balls you drop.[5] And even people with a partner or housemate who will pick up the slack may want to take a different tack. Here is the beauty of stickiness. Stickiness is the problem but it can also be the solution. Or at least it's part of the story for both.

Pam and Jason, a couple I know, created admin pathways that went against the social grain to share their parenting labor. Jason has the kind of job that means he can't leave work for face-time parenting—school pickups and sick-kid emergencies—but Pam can. To spread the job of parenting around, they decided he would do anything that could be done in the middle of the night. So he not only pays the school bill, he arranges playdates. Pam gets a sense of how unusual this is from how effusively the other mothers praise Jason. Interestingly, Pam also notes an efficiency improvement when Jason represents the household, since other parents seem more respectful of his time—an observation reminiscent of a famous story about Ruth Bader Ginsburg's parenting, back when she was a leading civil rights lawyer and a professor. Her son James, Ginsburg has said, was "what his teachers called 'hyperactive' and I called 'lively,'" so his school would call often.[6] One such call came after Ginsburg had been up all night writing a legal brief. She picked up the phone in her office and said, "This child has two parents. Please alternate calls. It's his father's turn." After that, according to Ginsburg, the calls came barely once a semester because "they had to think long and hard before asking a man to take time out of his work day to come to the school."

For the same reasons that admin often stays where it first lands, admin can also be redirected to places you choose. Admin typically involves pathways of interactions and information. If you reroute those pathways, you can make a new path that sticks. Some new pathways will work better than others. Some systems won't allow you to make the switch. But where you succeed, stickiness can become your ally.

Now that we've explored the costs, the gender dynamics, the imbalances, and the stickiness of the problem, are you feeling overwhelmed? Describing a problem in several dimensions can make it sound bigger than ever. As one woman asked in her brainstorming session, "Can we pay attention to [admin] and make it smaller?"

Don't worry. The rest of the book turns toward surprises and solutions. You can change. Your partner and friends can change. Admin itself can change.

Part II

# admin surprises

# 6

# Admin That Can Wreck You

Dealing with all of this "real-world," official, and essentially bureaucratic stuff combined two of my least-favorite feelings:

If you can pass the job on to someone else, I'd recommend it. If not, you have my total sympathy.

—Roz Chast, *Can't We Talk About Something More Pleasant?*[1]

Some admin relates to events so painful or scary that the admin itself becomes excruciating.

Death. Serious illness. Infertility. Bankruptcy. Divorce. Incarceration. Losing your home. These things may be happening to you or to someone you love or for whom you feel responsible. A parent. A child. A partner. A sibling. A dear friend.

Any list of events that engender catastrophic admin will include too much and too little. The reality of what can crush us is highly individualized.

From what I have seen, though, the strongest contenders for true admin wreckage involve three things: intense pain, uncertainty, and some need for action. The real or likely events are inherently painful. You can't control the course of events, and the right course of action isn't obvious. Nonetheless, you have to make decisions along the way.

Your father is unable to take care of himself but doesn't want to move out of his home into assisted living. You have frightening symptoms, like dizziness and fainting, but your doctor cannot identify the cause.

This chapter focuses on the worst of the admin of life. Anyone can hit an admin onslaught triggered by a sudden life event. But certain circumstances may make you particularly vulnerable. We'll start with the Sandwichers.

## Living in a Sandwich

Many of us trudge through admin onslaughts. Sometimes, for some of us, heavy admin burdens come from multiple directions, and we just keep soldiering on. We "do what has to be done, again and again," like the people celebrated by the poet Marge Piercy.[2]

The Sandwich Generation—those adults just the right age to be caring for declining parents while their own children still depend on them—are squeezed on two sides.[3] The so-called Club Sandwich involves triple-decker caretaking: for grandparents, parents, and children; or, alternatively, for parents, children, and grandchildren. All these layers of caretaking require a Sandwicher to become an expert in diverse kinds of admin.

The admin at the top of the sandwich—caring for aging relatives—can involve doing government paperwork to access public benefits, private paperwork to process insurance claims, or both; researching retirement facilities and managing an unwanted move; or taking over the admin for someone who can't handle it anymore. These examples barely begin to describe the labor of this care-work—especially if the aging relatives confront any especially challenging ailments. It was the stressful slog of applying for government benefits for her ailing parents that inspired Roz Chast to draw the cartoon that begins this chapter.

Research suggests that women do more care-work for aging family members than men and that intergenerational care is more common in ethnic minority families than in white families in the United States.[4] But anyone can be squished by this work, whether the direct care or the admin or, often, the two together. And the "emotion work" involved can be intense—whether a person is trying to rein in negative emotions or encourage positive emotions toward the relevant family member—which can make dealing with the admin even more exhausting.[5]

Similar challenges face the bottom of the Sandwich—raising kids. Under ordinary circumstances, kids leave a trail of admin in their wake. (I've come to think of it like the whirl of mud that surrounds Pig Pen in the *Peanuts* cartoon.) If kids also have disabilities or any kind of complicated health issues, the admin can be more like an avalanche. The work may involve researching possible diagnoses and schools, finding apt professionals, completing endless forms, reading reports that emphasize your children's weaknesses, attending grueling meetings, and, in some cases, filing a lawsuit. As Rachel Adams writes about caring for her son with Down syndrome, "Just days after Henry's services were approved, the therapists started to arrive. . . . Suddenly, I was in charge of finding, scheduling, and interacting with an entire staff of caregivers."[6] This is in addition to all the usual kid admin.

My interviewee Nancy hit the Sandwich jackpot. Both her mother and her son have disabilities requiring substantial research and care. Her mother has a rare condition similar to Alzheimer's that makes self-care very difficult, leading to health and hygiene consequences. Nancy worked hard to find her mother an excellent assisted-living center, but her mother still needs an advocate—to educate the staff about her rare condition; to find doctors to diagnose and treat her secondary ailments; and, these days, to tackle smaller issues that arise. The latest example: Residents at the assisted-living center are shown a movie in the evening, but, Nancy's mother tells her, "I never get to finish the movie." Staff members have scheduled washing her and putting her to bed at a time that conflicts with the nightly movie. Hearing her mother's concern, Nancy thinks, *Okay, how am I going to deal with that?*

Nancy's school-age son has dyslexia. Nancy spent a long time getting to a diagnosis for him. No one in his school seemed to understand the condition, so Nancy had to use her own time and resources to find specialists and get him evaluated to figure out the cause of his struggles and then, ultimately, to identify a good school for him. Now that his school matches his learning style, he needs at least an hour less sleep each night. (Previously, school exhausted him utterly.)

Though she's happy with his school, her admin continues. She believes that public-school teachers should be equipped to screen for, diagnose, and respond appropriately to dyslexia so no parent has to do what she did. Nancy is therefore building an organization to advocate for change in the public schools. She has become an admin crusader on behalf of other parents.

This is Nancy's way. "Organizational stuff is a strength" of hers, and she exercises it generously. She's the kind of person who tells you when there's a block party on *your* block. After she put her son's swim-team schedule on Google Calendar, she sent it to all the swimmers' parents. A natural Super Doer, Nancy believes technology has been helpful for people like her "who are truly admin people," but not for most people. (These days, because of email, she sagely observes, everyone has "their own admin job.") Nancy happily passes on the benefits of her tech-savvy admin doing.

Most of us can't do what Nancy does. Most of us facing admin that can wreck us are just trying to find a way to survive. But Nancy's story highlights one key Admin Surprise: No matter what you are facing, others have gone before you, and odds are pretty good that you can find someone among them who really wants to help you. And these admin nodes might even enjoy sharing their insights, making broader use of the relevant information filling their heads or the strategies they've devised along their own admin journeys.



## Losing and Finding and Losing Your Home

So much admin of poverty is admin that can wreck you.

Alexis, a nurse and mother of four, offered stories from when she used to live in federally subsidized housing. She described "the long process of getting a house approved" after "the long process of waiting for your application to get accepted." The process includes many steps and can go awry at any point for arbitrary or discriminatory or just plain bizarre reasons.[7] And "all that time, whatever living conditions you're trying to get away from, you're stuck there until they approve" your new place. Once Alexis lost her housing because the place had spiders—"Not regular spiders; it was in a wooded area" and they were "weird ones; like a white one, or a big one, or a fuzzy one"—and she complained to the landlord. The landlord wouldn't pay for an exterminator and so the place failed the federal standards for subsidized housing and she lost her home.

Alexis now lives in low-income private housing, and her housing admin is still invasive. When her twins were born in the month before our interview, one of Alexis's first admin tasks was to modify her lease. "Within a certain number of days of them being delivered, you are supposed to give proof of birth." Her children were added to the lease initially as "unborn 1 and unborn 2," and in lieu of a birth certificate, the landlord accepted a copy of her credit card. Needless to say, wealthy people don't have *send my credit-card info to the landlord to get the new baby on the lease* on their new-baby to-do lists.

## My Own Private Admin Hell, or How Admin That Can Wreck You Can Also Force Change for the Better

In the summer of 2015, I was staring down a dark tunnel of admin. I was separating from my wife, preparing to move across state lines, transitioning both of my children (then ages five and two) into new schools, and facing the reality of financial debts that were the shrapnel of a failed partnership. I was waking up at four a.m., unable to fall back asleep.

I was terrified.

I knew people who had been wrecked by their divorces. Most everyone does. Their eyes vacant, their hair thin and wiry, they look like patients near the end of terminal illnesses. I had always assumed the reason was emotional suffering.

As that summer got under way, I began to suspect the cause was more complex: emotional pain mixed in with a form of work that can capture your mind and your time, tracking you down at all hours and holding you hostage to anxiety and stress and the full panoply of negative feelings associated with the underlying event. In other words, excruciating emotion plus admin onslaught equals soul-crushing hell. Or that's what it felt like at the time.

I was plagued by the question of how I would get through this period—how I would do what needed to get done, for my children especially, and not lose myself entirely in the process. Scheduling movers, tiresome in regular life, now carried with it wrenching interactions and feelings of loss. The mere prospect of reading through a sample "parenting plan"—an example our mediator gave us of how some anonymous divorcing people had decided which parent would have the kids which days each month, each vacation, each holiday—stopped my breathing.

Every task felt overwhelming in its own right. To contemplate the full picture of what needed doing was impossible.

So I tried not to take in the full picture. I tried to find refuge in process. I had been running a number of admin brainstorming sessions in recent months, including with staff of a nonprofit focused on poverty. This group sparked my main innovation of this period, what would later come to be known as Admin Study Hall.





## I'll Have What They're Having

The mission of the nonprofit was to help low-income people seek and retain housing, jobs, and public benefits. The frontline staff who came to my brainstorming sessions were highly motivated volunteer interns and paid fellows. They were earnest and caring. Initially, they were unsure if spending two afternoons with me, instead of with their clients, was the best use of their time. But they showed up, played along, and quickly began to learn from one another.

Two staffers—Katrina and Susanna—commiserated about a client who would come in with a big bag of her unopened mail. This client would plop her mail on the desk and want to spend her appointment just going through it. For Katrina and Susanna, this did not feel like a good use of their time and training. The staffers were versed in navigating complicated benefits regimes; they knew the best ways to prepare a résumé and conduct a job search. They wanted to make the most of what they knew.

Katrina and Susanna laughed a little when they first talked about this woman—not unkindly, merely out of frustration. But as our discussions of admin progressed, their attitude to her shifted.

Talking about the staffers' own admin was critical. Our first session began with their selecting images from a sea of photo cards. They each chose one image that reflected how their admin *currently felt*; the group's choices included dominoes, a climber scaling a steep rock face, a messy room, and a crying child (which made everyone laugh). They then chose an image that represented how they *wished* admin would feel, and here, the selected images included a calm garden path, soldiers in orderly lines marching to war, and a bungee jumper who looked like she was having a good time. Several staffers sounded frustrated or baffled by their admin; two sounded very much on top of things. A texture was emerging.

Then we turned to their clients' admin. Partway through, Katrina suddenly connected her own admin failures and her clients'. She looked at the group, her face frozen with insight: "I give out a lot of advice I don't follow."

The next week, we met again. Katrina had begun to make changes in her life. She had implemented a new approach to all personal tasks that needed doing—both admin and chores—which was to ask herself, *Why not do it now?* Her habit had been to put things off, necessary things, and she wanted to change that feeling of running behind. She didn't force a new regime, but she put gentle pressure in that direction. A light contentment crossed her face when she talked about the new feeling of order in her world.

And yet she knew her admin obligations were mild to medium at this phase in her life. No dependents, no need to support herself, even, and few things to manage other than her own time and preparations for an upcoming semester abroad. But she was also undergoing a major transition, as were most of these young people, from her parents' doing all of this work to trying to figure it out for herself.

Focusing on her own admin made her look at her clients differently. Katrina no longer laughed about the woman with the bag of mail. She seemed to understand why the woman might want company facing it.

As we talked, I also started to take this woman seriously. For someone in poverty, opening her mail might mean finding something awful—awful news about her admin or awful admin that needs doing. I began to think that, like the woman with the bag of mail, perhaps what many of the nonprofit's clients needed was company. Possibly as much as they needed intensive vocational or benefits assistance, these individuals needed structure and companionship while doing their admin.



As the next few months brought me tsunamis of painful admin, I realized something else. I needed admin company too.

## Admin Study Hall

I began to imagine a space where people got together to do their admin. Side by side. Silently or with dialogue. Supportive interactions carefully and intentionally incorporated before, after, or even during.

My other main brainstorming group of this period—with members from a women's gym—culminated in a session where everyone brought in some of their admin to do. Lauren called it Admin Study Hall. The name stuck.

I began to incorporate Admin Study Halls into my own life. I had at least an hour a day of pressing admin during this period. So did several of my friends scattered around the world. We would do admin dates—mostly by videoconference to manage geographic distance or just the reality of kids or limited travel time—to force ourselves through some admin. These meetings had the added bonus of allowing us to see friendly faces, and we sometimes took a few minutes afterward to virtually clink our very real wineglasses.

What is helpful about Admin Study Hall (or ASH, for short)? Why bother to include other people? Participants say it makes the admin seem less overwhelming at other times because you know there's a window coming when you will definitely tackle it. Some say that talking about admin and doing it in parallel makes the admin more real, and thus less demoralizing when it takes way longer than expected, since someone else is having that experience too. ASH can also help support a feeling of accomplishment when you finish something taxing and hear "Congratulations."

The most emphatic comment about Admin Study Hall came from two interviewees who independently described the concept as "revolutionary." Their reason: ASH turns an experience that is usually isolating and even lonely into something collaborative, supportive, social. To have someone there with you, or more than one person, struggling through their own admin, may make real the message "You are not alone."[8] The need for connection may be deeply human, and so admin company could be helpful anytime in life, but in a painful admin onslaught, that sense of community can be an exquisite relief. Plus in ASH you know you're supporting another person's life as well.

A tight structure can keep Admin Study Hall from turning into Admin Social Hall. The basic model has four steps, outlined in figure 3.

Figure 3

**THE 4 STEPS OF AN ADMIN STUDY HALL**

1. *Brief check-in.* The shortest version is to each say one word about how you're feeling.
2. *Announcing intentions.* You each say what admin you're planning to tackle and what rewards you'll give yourself for the effort.
3. *Work time.* Work silently alongside one another—in reality or virtually—for the specified time; say, thirty minutes. (You may want to agree at the start whether to stay silent or interrupt as needed for moral support and whether to hit the "snooze button" for more time or to end promptly.)
4. *Checkout.* You each report on whatever got done, celebrate or complain about the process, and hold one another accountable for delivering rewards.

Many variations are possible depending on who's present. With my close friends, we often want to start by connecting more fully but still briefly, so we might describe the best and worst in our lives right now. In some settings, people I know like to begin with a short meditation or to end with a poem or a drink. I often try to build in what I've come to call Thanking God admin. By this (rather embarrassing) term, I mean admin that tries to do something good. It can be as simple as emailing a friend the name of a health-care professional or sending an appreciative note. Since this feels good, it's like an admin reward for doing admin. That and some dark chocolate (my usual fun reward) helps me leave ASH more prepared to return.

## Admin Shifting, Sharing, and Supervising

As I floated the idea of Admin Study Halls to various friends, I came to understand that, for some, the idea of doing admin alongside others ranged from not-worth-the-trouble to anathema. I also realized, over time, that Admin Study Halls involve coordination —the admin of setting them up, on top of finding someone else who wants to do admin together.

My darkest admin days brought further innovations.

One such strategy came during the first wave of divorce admin when my friend Eric, who'd been through similar hell, asked if he could help. He probably meant that he was willing to listen to me complain. I had another idea.

> You were so great to offer to help if you could, and you may regret it, as I'm coming with kind of an odd request. Could you possibly electronically babysit me for like 30 minutes today at about 3:30?

I needed to read through a sample child-custody arrangement to get some idea of how to set out where the kids would be and when.

This was the anonymous document that would stop my breathing. With some caveats, I explained my request to my friend Eric:

> If you happen to have the time, all it would involve is emailing me at 3:30 and saying "ok, time to open the document and read through it" (that is, the document from the lawyer with the sample parenting plan agreement), and then check your email during that half hour in case I melt down or want you to look at something, and then at 4 pm write me again and say "ok, that's enough; you should stop looking now; there will be more time later; go to the gym now, or do some writing."

He was game. This got me through the first read of the document.

I began to see that there were lots of ways to support someone, or be supported, in doing admin. There was *Admin Shifting*, where someone else does your admin for you—or you do admin for someone else. (Multiple interviewees talked about doing this, or resenting or resisting doing this, for family or friends who'd moved abroad.) There was *Admin Sharing*, where someone joins you in your admin-doing. (My mother, bless her, traveled to another state to walk through thirteen rental-apartment prospects with me when I was separating and preparing for an interstate move.) This exchange with my friend Eric added to the list *Admin Babysitting*.

Admin Babysitting—or as I call it when I want to feel more grown up, Admin Supervising—just involves someone encouraging you to do the dreaded admin by knowing when you're doing it and checking in on you occasionally, whether in person or remotely. This is distinct from Admin Study Hall, in that the babysitter isn't necessarily doing his own admin. He's just . . . there.

I realized recently that my father deserves distant credit for the Admin Supervising paradigm. In college, he once helped me get through my panic over a writing deadline by asking me to call him every hour and update him on my progress. By the end of the day,

I had a plan for the paper that would keep me awake through the night to write it.

Admin Supervising is probably best suited to extreme or emergency admin situations that justify the setup admin. That said, once the architecture is in place with one or more people, the startup costs for an individual session can be minimal. ("Admin-sit me from three to three thirty today, if you have time?") And sharing the topic of the admin is optional. Surely some would find this infantilizing, but when the admin is bad enough, childhood can look awfully appealing.

To the introverts among us, the social aspect of Admin Study Hall or even Admin Supervising may be precisely what ruins the idea. More appealing may be a variation on Gretchen Rubin's *power hour*. Rubin uses this catchy term to describe a once-a-week window for tackling projects that aren't time-sensitive and therefore might not ever get done because, as she shrewdly notes, "Something that can be done at *any* time is often done at *no* time."[9] In an admin crisis, the power hour (of mostly urgent items) may need to be once a *day*, or *many times a day*, not once a week.

This brings us to one further example of really awful admin that can wreck you.

## Admin No One Should Have to Do

Before going on a cruise with her husband a few years ago, Kimberly Norwood, mother of four, had a special form of admin on her plate. "At the top of a long list of things to do before we left for our trip was 'email chief of police,'" she writes.[10]

Kimberly sent the police chief pictures of her sons and information about their route walking to summer-enrichment classes at a school a mile from their home. She was not worried about abduction or child molesters. She was worried about protecting her sons from the police. Her sons are black. In their suburb, twelve miles from Ferguson, Missouri, 3.5% of the people are black. Kimberly was concerned that someone might call the police on her teenage sons as they walked through their own neighborhood to summer school.

These fears are not hypothetical. Her teenage daughter has been stopped three times at the mall on suspicion of shoplifting, sometimes singled out in a group of white friends. (And always found innocent.) And when Kimberly is pulled over for speeding in the neighborhood—for going forty-five in a thirty-five-miles-per-hour zone—the first question is whether she lives "around here." These are just a subset of the experiences that put the police-chief email on her to-do list.

Her pre-cruise email is one example of the admin of discrimination, aka *bias admin*, which takes many forms. Bias admin is admin no one should have to do.

## Ending on a Happy Note

It's hard to go from deeply troubling admin—painful all the way down—to something lighter. But I promised you we would talk across the divides of quantity (inviting those not in onslaughts to relate to those inside them and vice versa) and privilege (remembering that we all suffer in different ways, even if we can agree that some ways are worse than others). So I'll conclude by noting that even happy events can generate admin that can wreck you, at least temporarily.

Such happy events may include buying a house, getting married, moving for an appealing new job or degree course, hosting a life-event party, even entertaining your parents visiting for graduation—and, of course, having a baby. They may not involve the painful layer of heartbreak or fear entailed by divorce or illness or discrimination, but they can nonetheless take over your life for days or weeks or months.

Happy admin can be even more surprising than disaster admin. And admin for happy occasions isn't necessarily happy; the reasons for the admin might be pleasant, but being colonized by the planning might not bring the joy you'd hoped would come with the occasion.

Any of the strategies in this chapter, including study halls or supervision or admin nodes, can come in handy. The first step is recognizing an admin onslaught when you're in it—ideally before that, but at least during.

What is your admin quantity level right now? Has your admin crossed over into an onslaught? Or are you managing?

Perhaps you are even ready to consider how you might use admin rather than allowing admin to use you. The next chapter shows how to make that leap.

# 7

# Admin That Can Fix You

Keep in mind that you are always saying "no" to something.

—Stephen Covey[1]

What do you wish you could change about yourself?

Forget about admin for a moment and think broadly about your aspirations. Imagine it's one year from now. If you could have done any one thing this year to improve your mind or your body or how you spend your time—something challenging but possible, not requiring an actual miracle—what would it be?

Whatever the goal, admin offers an entry point to thinking about our unrealized aspirations and devising new plans for meeting them.

This chapter differs from what's come before. Rather than focusing on the problem of admin and how to fix it, this chapter aims to help you use admin to improve you. (Not that I think you need any improvement, but just in case you think you do.)

## Getting to Quadrant 2

My father loves Stephen Covey's *The 7 Habits of Highly Effective People*. So much so that I'll admit to avoiding this book for many years. Nothing against my father, who is smart and curious and has flagged many excellent books for me. But those of us who have been parents or children—that is to say, everyone—probably know the feeling of exhaustion from a close relative's effusive recommendations.

I found my way to Covey's *7 Habits* only as an adult, when a person I admire recommended his *urgent versus important* matrix, a variation on which appears in figure 4:

Figure 4

**URGENT VS. IMPORTANT MATRIX[2]**

| | Urgent | Not Urgent |
|---|---|---|
| **Important** | Quadrant I<br>• Crisis<br>• Pressing problems<br>• Deadline-driven projects | Quadrant II<br>• Relationship building<br>• Finding new opportunities<br>• Long-term planning<br>• Preventive activities<br>• Personal growth<br>• Recreation |
| **Not Important** | Quadrant III<br>• Interruptions<br>• Some emails, calls, meetings<br>• Popular activities<br>• Proximate demands | Quadrant IV<br>• Trivia, busy work<br>• Time wasters<br>• Many calls and emails |

Whatever your views on what belongs in each quadrant, the focal point of the matrix is quadrant 2, where *important* meets *not urgent.* Herein lie so many of our failed projects as well as our biggest challenges.

The fact that writing typically lives in quadrant 2 is the reason I cannot write in my office at work. There I am plagued—in reality or in my mind—by all the demands on me that come from official askers: students whose papers need grading, colleagues whose committees require attention, prospective employers who call about my former students, and more.

Writing is essential to my job, to my career, and to myself. But my kind of writing rarely has an insistent asker. On any given day, it is far too easy to feel I must begin my work with all those other demands that can seem more important or more urgent because someone else is requesting that I do them. And yet it will never be possible to get through all those demands in a day. And so, if I began every day by responding to the external demands, intending to complete them before moving on to writing, I would never write.

Many of our important endeavors are like this. You surely have one or more that fall into this category. "Keep in mind," Covey tells us, "that you are always saying 'no' to something. If it isn't to the apparent, urgent things in your life, it is probably to the more fundamental, highly important things. Even when the urgent is good, the good can keep you from your best, keep you from your unique contribution, if you let it."[3] This is the main insight of Covey's well-known matrix.

What matters to achieving our goals, whether as professionals or as humans, is often important but not urgent. Yet what occupies some other quadrants may be important to our values. Making calls on behalf of others—whether for my job or in the rest of life—will always be high on my priority list, for instance.

To achieve broader goals, we have to find a way to meet—but contain—what lies in the other quadrants, at least some of the time, to make room for quadrant 2.

## Asking the Admin Question in Pursuit of a Goal

This is the all-important Admin Question: *What role does admin play, or could admin play, in a problem or its solution?* Asking the Admin Question can help with any goal. Here are seven ways to use admin insights in pursuit of a life change.

1. ***Opting for low-admin means to your end.*** Different paths to a goal may involve more or less admin. For instance, playing tennis, which requires court times and partners, is a more admin-intensive exercise than walking or jogging. Indeed, any activity that involves other people—like starting a band or playing chamber music—likely entails more admin than an activity you do alone. Knowing this doesn't mean that you're wrong to choose the admin-heavy activity. But you need to take admin intensity into account lest it become an invisible drag.

2. ***Paving the way.*** Doing some upfront admin—like scheduling events in one's own calendar or with someone else—can create a path to keeping up with a regular activity. That upfront admin can reduce admin later and even remove the decision moment. For instance, signing up for a weekly language class could free you from repeatedly having to choose whether and when to practice the language each week.

   Scheduling activities that matter to us may make us happier. A recent study of women's time use found that women who schedule their free time regularly are more likely to report being satisfied with their lives than women who wait until they finish other tasks.[4] Though the study focused on women, it seems plausible that the same principle applies to everyone.

3. ***Picking partners wisely.*** If you do launch a collaborative endeavor, like playing a sport with a friend rather than running

alone, think about the person you're choosing. Does this person fail to show up or never return emails? If so, this person might not be a promising partner for coordination. Or is he someone who does a great job organizing everyone for pickup basketball and who will make your participation easy? A good admin partner need not be good at admin. Is her office right next to yours so it's simple to go to the gym together spontaneously? That may be enough.

Consider how easy this person is to coordinate with for *you.* If you want and need someone to organize you, then pick a Super Doer collaborator. If highly organized people annoy you, and their email reminders could drive you to rebel against the exercise regime you wanted, then pick someone more your style and be admin-savvy about *how* you plan with this particular person. Consider making a default plan—every Tuesday at lunchtime, say; if someone cancels, that person is in charge of proposing a time to reschedule or just kicking it back to the next Tuesday. Or with someone who cancels a lot, try to arrange things so your workout proceeds even when your partner's a no-show.

4. ***Consult nearby experts.*** Who in your world has successfully done what you're trying to do? Or researched and read everything out there on a relevant topic? As we saw with admin onslaughts, here too finding admin nodes can be a lifeline. Identifying a knowledgeable human whom you can ask the questions most relevant to your life—*What is the best exercise for someone with too much energy and joints that hyperextend? What's the best art studio for adults?*—involves much less admin than hunting around for information on a subject about which you know little.

5. ***Use admin as "sludge" to keep you on track.*** Gyms and cell-phone providers know that if they make canceling your

contract heavy on admin, you'll be more likely to stay. Companies issue rebates rather than just giving you a discount because they expect that most people will fail to do the rebate admin. These entities are using sludge to serve their interests. *Sludge* is a term coined by the Nobel Prize winner Richard Thaler for what might be understood as the converse of what he and Cass Sunstein call *nudges*—small actions, informed by behavioral science, that can have a big impact.[5] Sludge, by contrast, is when an entity makes certain actions very difficult through small burdens.[6]

The logic of sludge can be used for good as well as evil. Increasing admin drag on the back end of good choices or on the front end of bad choices—that is, imposing some sludge—can help redirect your choices or deter quitting. To promote healthier eating, for example, Google's headquarters replaced its "self-serve M&M station with individually wrapped packages and placed cookies and crackers further from the beverage stations."[7] Likewise, if you make a healthy restaurant your default location for your weekly lunch with a friend, and you have to send your friend new details if you want to change the restaurant, that's gentle sludge working for you. If you want to use your phone less, you could try creating a longer password and disabling touch ID. (This would drive me to distraction, but it might work for you.) Find ways to get sludge on your side.

6. ***Choose the launch date carefully.*** We generally can see that a new activity—cooking, exercising, gardening, paying closer attention to finances, whatever it might be—will take time and energy. We rarely appreciate the time and mental energy of the *admin* required to alter life around that activity.

   People often start new habits at symbolic calendar dates—like New Year's or a birthday—because these "fresh start" moments can help spur motivation.[8] But an admin perspective recommends thinking carefully about when and how you'll be able to manage a change of habits. For instance, starting a new diet may involve identifying apt recipes, finding new places to dine out or shop, learning to keep track of your eating for the diets that require this—all things that take time and mental energy. Recognizing these admin costs might tempt you to start a new diet on vacation; hence the appeal of spa retreats for healthy living. But if you start a diet during regular life instead, all that new information about where and how to get the right food will be right at hand, sparing you the admin of transferring the results to your home routine just as your days become busy again. So you may want to carve out the time in daily life rather than waiting for vacation. Or consider a staycation for your next break.

7. ***Cut out something else.*** Time for the admin of creating a new habit—figuring out the when and where and how—has to come from somewhere. What can you zero-out during the phase-in period? Put strict limits on email or media usage. Ask a family member to plan the meals. Think about where you can cut corners temporarily, until your bandwidth is no longer taken up by the admin of your change.

None of these points leads ineluctably to a particular path. Your circumstances and tastes are your own. The point is that asking the Admin Question can be a vital component to creating a regimen that works.

## Putting It Together

I saw these principles at work after my second child was born. With my first, the baby weight steadily fell off until I was nearly back to my pre-pregnancy self. Not so with the second. The weight initially went down and down and down, just more slowly. But then it plateaued. And after my son's second birthday, I realized things were reversing course. As my weight crept back up toward my pregnancy peak, I had to face facts.

After a month of debating whether I liked food more than I cared about my weight, I finally admitted that I cared. I wanted to lose the baby weight. And I wanted to feel healthier again. An admin perspective helped me figure out what had happened, so I could make a plan for change.

As a former runner, I used to be suspicious of group exercise. I loved that I could run almost anytime and anywhere, with little or no planning or coordination or expense. Low-admin exercise, I'd say now. This changed in my thirties. Perhaps it was the influx of external demands that came with that phase of life for me (colleagues, children, and the like); perhaps it was just becoming older and less wedded to a belief in my own autonomy. Whatever the reason, I found new appeal in the idea of classes when I could possibly manage them. What could be better than being shut in a room and forced to turn off my phone and do what the teacher said?

More than that, classes offered the opportunity to schedule far in advance (point #2, paving the way). I could choose the best classes for me (yoga, in recent years) and slot them into my calendar on repeat for the coming months. Then I would try to plan other commitments around those classes, leaving the option of scheduling over them as a last resort. (These days I frequent a studio that's by donation, a trend that is making yoga classes affordable for many more people.)

The admin advantage of my yoga regime had turned against me when I moved to a new city just days before starting my new job. I hadn't made time for the admin of locating the right studio and class times and slotting them in to my schedule. And so I attended almost no yoga classes those first few months.

After my son's second birthday and my realization that my waistline was growing, I used the Admin Question to plan my intervention. I designed a regimen of diet and physical activity that involved some admin startup costs I could pay during a relatively slow period of life and work, around New Year's (#6: choose a launch date carefully), and that would be low-admin in application once life got busy again.[9]

I joined a yoga studio with a monthly membership so the cost was the same no matter how often I went, and I booked myself for daily slots (#2: paving the way). I got some friends to join the same studio, but knowing our personalities and busy lives (#3: picking partners wisely), I merely told them about classes I was attending but didn't count on their participation. (If I'd made a firmer commitment to be at certain classes each week, then I could also have used the pressure of sludge (#5)—that is, of having to notify them if I bailed, whether or not they were coming—to help keep me on track.)

And I opted to cut out carbs, following my then-wife's inspiring example. As I'd learned by watching her, a low-carb diet has the appeal of a rule that you don't have to think much about, once you know which no-carb foods you like and where to find them (#1: opt for low-admin solutions). Adopting this diet also allowed me to free-ride on all the information she had gathered about foods, including some that were already in our home (#4: consult nearby experts).

To carve out the time, I tried reducing my teaching preparation for classes I had taught many times before (#7: cutting out something else)—so I could go to yoga in the morning before class. Since yoga improves my focus and mental clarity (a big reason I do it), I generally found that my prep was more efficient, and my teaching was more tuned in to the students.

Four months after beginning this new regime, I was close to my pre-baby weight again, had a regular yoga practice, and felt much better. Part of this happy result surely stemmed from luck and privilege. For most goals, most of the time, success doesn't come easily or quickly—particularly for something like weight loss, which can involve many complicating factors and obstacles. The point here is only that asking the Admin Question can be one tool to give us an extra advantage, some wind at our backs. And if we don't pay attention to it, admin can be a roadblock that thwarts even the best intentions.

When I moved again two months later, I had learned my lesson. I did my best to block out time to set up my habits before diving into work.

## Turning ASH into ISH

The summer of my second move brought another innovation. Admin Study Hall took admin to set up, and I began to realize that for that particular life moment, the setup admin was more worth it if the Study Hall carved out a space *without admin.*

With some willing collaborators I morphed Admin Study Hall (ASH) into Intentional Study Hall (ISH). With ISH, I could use the "gentle accountability" of the study-hall structure to tackle my highest priority item, something important but not urgent. For me that something was often writing.

For my ISH compatriots, important-but-not-urgent somethings ranged from preparing for immediate work events to pursuing a long-term creative project to searching for a job. ISH can be used for anything. The limits are set only by one's intentions.

I also created a template for a solo ISH so I could lean on the structure without the setup admin when I didn't need, or didn't have, a fellow study-haller. The basic solo ISH looks pretty much like the four-step process outlined in figure 3 for a Study Hall with someone else, except the chart and the clock are your companions.

Figure 5

**CHART FOR A SOLO INTENTIONAL STUDY HALL (ISH)**

| Minute on the clock | Activity for this step | Entries |
|---|---|---|
| :00—:01 | 1. *Brief check-in.* Write a word or two about how you're feeling. | ______ ______ ______ |
| :01—:03 | 2. *Setting intentions.* Write down your intentions for what you'll do in the session and what reward you'll give yourself after. | ______ ______ ______ ______ ______ |
| :03—:28 | 3. *Work time.* Do what you intended to do. | |
| :28—:30 | 4. *Checkout.* Administer reward; perhaps write a word of reflection or affirmation. (What do you wish someone would say to you right now?) | ______ ______ ______ ______ ______ ______ |

Start a stopwatch so you can see where you are in time for each step, and you're off.

## The Dangers of Being Fixed by Admin

Admin can also help fix us when we are feeling broken—at least temporarily.

People have described to me how, in the days after a loved one's death, they found making funeral arrangements a reprieve of sorts. The admin onslaught gave purpose and direction to a moment of emotional unmooring. A period of intense doing to hold emotions at bay may be useful, functional, when the feelings are too much to handle.

But at other times, admin may be protecting us from a vulnerability that's valuable.

Vera, our recovering Denier, offered me insight into this when describing her wedding planning. She and Saul had chosen to save time by doing things last minute. She described her frustration with trying to get rings and a dress a month before the wedding. Her calls were met with responses like "Oooh! Is this a quickie thing?" Which didn't make her feel very good, not least because it wasn't true. Wedding vendors expected them to be planning their wedding for next summer, not this summer.

Why do you think people plan their weddings so far in advance? I asked her. She was intrigued by the question. "I realize that I don't need some idealized version of the perfect wedding dress that captures my personality . . . just perfectly," she reflected. "I just need something that is good enough." (This made me think of psychoanalyst D. W. Winnicott's "good-enough" mother, credited with helping her child learn to tolerate frustration and disappointment by failing to do everything perfectly.) Then Vera made a broader observation. "Some people like to have some control, or just like to have all these details locked down way far in advance and then they can relax." She paused. "I think Saul is a little bit more like that than I am," she added.

Vera's insight resonated with what I'd been learning. People sometimes do admin far in advance not only because they care about getting it right, and not only for the feeling of control it gives them to have it done; people sometimes do admin far in advance out of a belief that they must get the admin done first, before they can relax.

After I spoke to Vera, I started looking for that belief in my own thinking. How often did I think that I'd relax right after I got this one further thing done? It's like waiting to pee until the whole house is clean. Except that, for most people, the house does eventually reach a clean-enough state—or the need to pee eventually sends them racing to the toilet. But as Vera also pointed out, the admin will never get done. The list will never end. And relaxation can't force an entry, except through sleep, which is surely the poor man's relaxation. (And for many of us, even the escape of sleep is interrupted by the four a.m. admin wake-up call.)

During trying times, the impulse to do admin may be even stronger. Death, divorce, illness of a loved one: all of these may fuel a desire for control and a sense that we can't let up until it's all done, and done well. We also may fear what we might feel if we stopped doing, if we turned off the mad spin of the admin hyperdrive, the *madmin* mind. After talking with Vera, I began to ask myself, *What would I have to feel right now if I weren't zooming through this admin, turning it over and over in my head, trying to get it just right?*[10]

One evening, my younger child sat on my lap crying about a favorite stuffed animal he'd lost several months earlier. I had replaced it with the closest approximation known to Amazon, but the new one is never the same. Time had passed, something had reminded him, and my child was having a good cry about missing the lost lovey.

I held him. I hugged him. I let him cry it out. Then I offered a suggestion—did he want me to make a photo book, a little set of pictures of his lost friend? He said no. I heard him. I said, "Of course not; of course you don't want pictures of Bowwow. You want Bowwow."

But I also felt a tug to solve the problem, to start the admin engine down the tracks. I wanted to find old pictures, upload them to

Shutterfly, have them ready for the next time. Perhaps his sadness felt like too much; I wanted somehow to save us both.

In the end, I compromised. I pulled up a few pictures while he sat on my lap, and I kept hugging him. In retrospect, I wish I'd just hugged him with both arms.

8

# Admin Judgments

Have you ever noticed when you're driving that anyone who's driving slower than you is an idiot, and anyone driving faster than you is a maniac?

—George Carlin, from the stand-up routine "Stuff on Driving"

Admin is like driving in the old George Carlin joke. Everyone else is going too fast or too slow, doing too much or too little, being too anxious or too lax. Everyone else is a maniac or an idiot. Which means, of course, that to all the idiots you're a maniac, and to all the maniacs you're an idiot.[1]

I asked my interviewees if they knew anyone who did too much admin or too little admin. There was often an excited energy, even a frisson, in their answers. Several people said they just didn't know enough about other people's admin-doing to answer the question. But most people immediately thought of someone to judge in each direction. Often the person who came to mind was a sibling, a child, a parent, or another relative. Sometimes a friend. Or a coworker.

Colin, who is British and quite enjoys some kinds of admin—like hunting for better deals on his utilities or complaining about faulty service and seeking compensation—thought of two of his female friends. Both women declined invitations to go to the pub sometimes because they needed to stay home and "do their paperwork," he said. (Colin acknowledged the possibility that the women were making excuses not to go out with him, but, as a happily married gay man, he was well understood not to be looking for more than their friendship.) He realized he too could benefit from spending more time filing his paperwork, which he dumped into a box. If he organized his papers, then when he occasionally needed a particular paper, he wouldn't have to spend time hunting. But he generally felt pretty comfortable with his decision to go to the pub instead. He didn't want to be that person who spends his spare hours filing papers just to avoid the later hunting.

Chloe expressed anger toward the maniacs. When I first asked, she tried to be tolerant, describing theirs as "just a personality type." But when she thought hard about what advice she would give the people who did too much admin, her energy increased: "My advice would be back the fuck off . . . There is also kind of a 'I'm more admin-y than thou' superiority complex that those people like to throw around."

The maniacs also have some harsh words for the idiots. *Flaky* is the most common epithet.

Cheryl, who forms a Super Doer trio with her politician husband and their assistant, Ellen, put the point more philosophically. When I asked if she knew anyone who didn't do enough admin, she said, "Probably many people." She continued, "Woody Allen said it best: 'Most of life is showing up.' But if you're not organized enough to figure out how and when to show up, then forget it; you're behind to begin with."[2]

## Don't Zen on Me

I have felt admin judgment from both directions. The place where I've most often felt judged as an admin maniac is in the world of meditation and yoga.

I bring up this topic with some hesitation. Meditation has "a towering PR problem," to quote the ABC News anchor Dan Harris, so for those who don't practice it, this brief discussion may be, at best, boring and, at worst, aversive.[3] And for those who do practice meditation or yoga, my describing my frustrations may seem ungrateful to communities that have given me and others so much.

I am grateful—and frustrated. It's fair to say this book wouldn't exist without my meditation and yoga practices. Meditation allowed me to see admin. When I sat in the corner of my bedroom after my second child was born and began making a list of all the things I was doing in my household—a foundational moment for my identifying admin as a thing—I was in the chair where I went to meditate. After kids, especially my second, meditating was so much less pleasant. The aim of mindfulness meditation—which involves noticing what's happening in each moment—is not to have a pleasant experience. But it sure is nicer to have pleasant experiences. Sitting in that chair, watching my mind, I had seen how filled it was with logistics, more than I had ever imagined possible. Meditation allowed me to see that and thus to have the idea that became this book.

And since then, yoga, even more than meditation, made it possible for me to *write* the book. Yoga is the best antidote I've found to madmin mind. People differ widely in what helps free their minds of admin, even for a moment, during an onslaught. But for me, yoga has been the most reliable path to clearing my mind enough to do anything requiring mental focus—including writing.

But yoga has not only been a solution. It has also been part of the problem. Yoga teachers will sometimes say, "Let go of whatever is on your mind. If it's important, it will be there when class is over." The teachers who say this are not really hoping that just moments after the final pose I'll be thinking, *Yes!* Now *is when I'll click on the Amazon app and order the diapers.* No, what they seem to mean is that the things on my to-do list just aren't that important. And that the yoga practice will help me see that. Similarly, meditation teachers sometimes invite you to ask yourself, "What if I died? Which of the things I feel I need to do would really need to be done after all?" When I first started meditating and I was relatively unencumbered, without a spouse or children, perhaps I found it liberating to ask this question. Perhaps I really did conclude, as the question urges us to, that most of my to-do list really wouldn't need to be done after all. That most of it was *just in my head.*[4]

Yet once I had people depending on me, many of the things on my list would remain urgent or important (or both), even if I died. My younger child would still need diapers ordered (or bought somehow); the rent would still need to be paid (so my children had somewhere to live); childcare would still need to be found or managed (probably more so in my absence); and distant projects like doing a will would seem even more consequential (although it would be too late, since I'd be dead). Asking "What if I died?" no longer made me feel that my to-do list was unimportant. And yet the message from some fellow meditators sometimes seemed to be just that.

I have felt similar frustrations around cell-phone rules. Meditation centers—like other venues that support going off the grid (such as some vacation sites) or that rely on silence or darkness (such as concerts or movies)—often ask participants to turn off their phones. This makes sense to me. *I want my phone off.* I want *other people's* phones off. I am well persuaded by the research suggesting that the bandwidth for complex tasks is decreased by the mere sight of a smartphone screen, even someone else's.[5] What I want is some way to be reachable through their front desk in the event of an emergency involving my kids—and occasionally, now, my parents. Even at well-funded institutions with support staff, these requests have often received no uptake. Organizers have sometimes seemed annoyed by the question, suggesting that the need to be reachable, like much admin, isn't being seen.[6]

Whatever we do to escape the grind of admin—and whatever else is grinding us down—may give us perspective on which tasks are important, which are urgent, and which are neither. But taking care of ourselves isn't going to get our important admin done. If it is March 31 and my pretax-dollar submissions on my health-care Flexible Spending Account are due today, and I have several hours of work to do in sorting and submitting them, then I cannot go to yoga during my only window of free time. No amount of yoga is going to make that FSA admin go away.

My ex-wife and I had a phrase we used to say when one of us, having just emerged happily from a yoga class, would try to shrug off—or judge—the other's stress. "Don't Zen on me," the one who'd been home with the kids would say. It's so easy when you're in the breezy, relaxed place to judge other people's admin-doing as maniacal. But please don't Zen on me. And I promise not to Zen on you. Or at least I'll do my best. Not judging other people's admin-doing or admin-not-doing is a lot harder than it sounds.

## Why Admin Is Ripe for Judgments

Have you ever noticed how, between two people, the smallest difference can become enormous? Two similar people can begin to feel like opposites. Comparative judgments are possible along any dimension of difference, yet admin seems particularly prone to elicit judgment, for reasons related to admin's particular features.

Admin's *invisibility* makes it easier to ignore the other person's efforts. Another person's time spent on the parallel shift may go unnoticed—or even seem implausible. For this reason, my interviewee Colin thought the occasional arguments with his husband, Adam, were useful to dispel comparative illusions: "It's kind of good to bring it up in some kind of annoyed moment and actually get told, get reminded" that the other person is doing this stuff too. "When you're in a couple," he added, "it's easy to forget that somebody else is taking care of some stuff, but of course it's not happening by magic."

The perception of admin as *trivial* can make it easy to dismiss other people's efforts, even when we do see them. This can fuel disputes over the value of admin, disputes that may be more robust, more extreme, than in other areas. Even if partners disagree about the *hows* of toilet cleaning or dishwashing—how often or how promptly or how well—it is hard to imagine either partner saying the job *never* needs to be done at all. The dirty toilet or dishes stare not only at you but at any visitors to your home. By contrast, one partner may really think that certain things are important—like researching restaurants before choosing; or scanning and storing important documents—while the other person thinks they are an utter waste of time. This can make the judgments and negative emotions more intense.

There's a big difference between something being given a low value and something being given zero value—on both sides. Think how bad it feels to spend your time doing something and get no credit at all. And then think, from the other perspective, how bad it feels to be blamed for *not* spending time doing something you think is entirely unnecessary.[7]

Knottier still, there's the problem of the varying textures of admin. Different kinds of admin differ widely in the toll they take on your mind. Bombardment admin, which constantly interrupts you and demands immediate responses, impinges on your mental space and the feeling of your day, at home or at work, more than admin that is under your control, like paying monthly bills. But that experience is hard to appreciate if you're not living it, making it difficult to understand why you're being judged for not being the point person.

## Before We Judge

How can we turn admin judgments into admin compassion?

When I'm inclined to judge someone for non-doing, I think of

Chloe, who helped teach me about our collective interest in rebelling against admin impositions. This is what Chloe would like to say to some of the maniacs she knows: "The more work you do happily and compliantly, the more work it makes for everybody else." Chloe, a doctor, was talking about admin on the job, where her concerns about paperwork are not taken seriously if other people say the admin burdens are easy and nothing to complain about. Her point applies much more broadly, though. To take one example, I've heard many parents lament the endless back-to-school forms they face each fall (which are sometimes made better by technology, and sometimes made worse, depending on the technology). So long as all the parents just quietly complete the forms and submit them, a school may never know the burden it's imposing. Non-doing can serve the group as well as the individual.

And sometimes, admin non-doing, or doing badly, is the best someone can do on a given day. I have certainly had those days. When I see someone in one of those moments, I try to remember the sentiment—variously attributed to Plato, Pliny of Alexandria, and (the likelier prospect) John Watson—"Be kind, for everyone you meet is fighting a great battle."[8]

When I'm inclined to judge someone else's *doing*, when I think someone else is working too hard to get some admin right, I remind myself that I can get judged for my admin doing in the domains I care most about. Our particular values come through in what admin we do and how. Telling someone else, "Relax—it's just not that important," can sometimes be a way of saying, "What you care about is just not that important to *me*."

And if I were seeking sympathy for my being a Doer at all, from someone who looks down on the whole enterprise of doing admin, what then? Here, I'd invite into the conversation Chris, my proudest Avoider. When I asked him a question about the limits of outsourcing—"Are there any forms of admin you would not give up, even if you had the ideal robot or personal assistant to do them?" —Chris's first response was "If the robot did it perfectly, I'd let it do it all." But then he paused and reconsidered. Admin involves questions like "How much am I spending on groceries versus how much savings do I have?" Chris ultimately concluded he couldn't give it all away. "Because, in a sense, admin is this way in which you reflect on your life."

## Trust Others and Know Thyself

Lastly, let's not forget the prickliest form of judgment for many of us—self-judgment.

When I interviewed her, Avoider Diana lamented having given her mother, Pat, a ceiling fan and promising to have it installed—over a year earlier. (Three months after that, when I interviewed her mother, the fan was finally functional.) Had Diana remembered her admin personality, she might have hesitated before giving the gift of admin. Instead, she got a year's guilt plus a lingering to-do list item.

Know yourself. Trust others, where you can. And try to respect others and their admin ways. One way to start to develop that trust? Use the insight of idiots and maniacs to relativize your own position—to realize you play both roles, depending on your audience—and to learn from other people. There are a lot of idiots and a lot of maniacs out there.

9

# Admin Pleasures

There is a big secret about sex: most people don't like it.

—Leo Bersani[1]

Sometimes research confirms what you already know. Sometimes it surprises you. One surprise of my interviews and brainstorming sessions was a reversal of the classic line about sex from Leo Bersani. If the big secret about sex is that many people don't like it, then the big secret about admin is that many people do like it.

The crucial point here is about secrets and confessions. Bersani's line is surprising because we so widely assume that people like sex. And indeed, many people, much of the time, do like sex, as real-time happiness surveys confirm.[2] But most people, most of the time, also feel they *should* like sex. It's hard to admit to not liking sex. (If you don't believe me, ask an asexual person.[3])

We face a similar dynamic around admin, but flipped. The hard thing to admit about admin is *liking* it.

Person after person told me, almost confessionally, that they actually kind of enjoy admin. At least sometimes, under some circumstances. The surprise here was not the tiny number of people who emphatically love filling out forms. (What unusual pleasure doesn't inspire an internet fan site?) The surprise here was that ordinary people were confessing even mild enjoyment at something most of us typically dread and dismiss.

This observation should not be confused with people wanting to spend more time on admin or believing that it is a good use of their time—a topic we'll discuss. And no one claimed to actually enjoy doing particularly onerous or painful admin tasks, like searching for a nursing home for an ailing parent.

But if many of us have the potential to find even a small amount of pleasure in some admin, why don't we know this? What kinds of pleasures are there? And what can we do with them?

## Admin Confessions, or Repressing the Impulse to Admin

Zack observed that admin's not cool. And it's not sexy. He channeled a collective sentiment: "*Do we really have to talk about that shit?*"

In my interviews, those who admitted liking admin sounded as if they were sharing a guilty secret. Jane offered the most self-conscious confession. Asked how she felt about admin, she replied that she was "ostensibly resentful." Why ostensibly? Because she buys in to the rhetoric of "If only there weren't all this admin, I could be thinking great thoughts." In reality, though, she said, "I don't mind it nearly as much as I claim I mind it."

By noting the gap between what she said and what she felt, Jane was identifying something important: The pressure not to like admin can work to preserve our time for more important matters. Admin is dangerous. As with Covey's two not-important quadrants—quadrant 3 (urgent) and quadrant 4 (non-urgent)—admin threatens to disrupt our broader goals and priorities. Perhaps for this reason, many of us have built up a resistance to it, layering a special distaste or resentment on top of whatever feelings accompany the tasks themselves.

Not liking admin may be adaptive in the way that not learning to type was thought to be adaptive by those professional women before computers became widespread. Not liking admin may help protect us. And since so much admin comes from the outside world that's making demands upon us, don't we need that protection? Not liking admin is a useful heuristic, a rule of thumb, to conserve time and energy.

But admin-hating, like any rule of thumb, can help in many situations but misfire in others.[4] Our dislike and trivializing of admin, which helps us to deflect this work and protect our time and mental space, may disserve us when the admin really needs doing. Or when we are trying to give credit to others for their admin labors.

## Pleasurable Positions: Done-It and On-Top-of-It

So what do admin pleasures look like?

The most common form I've observed is what we might call *completion pleasure* or *done-it pleasure*. This is the satisfaction of crossing things off the list, of getting something done. And sometimes, the more difficult the task, the greater the relief when it's done. A triumphant glow may come with winning a battle with the recalcitrant seller of a defective product. But even striking a tiny item off the list can feel good. (Those of you who've created a list midway through a project and included items you've already done—just so you can cross them off—know what I mean.)

A close second for admin pleasures—probably second only because so few people can achieve it—is the feeling of having a good system for keeping track and staying organized. For some people I interviewed, the task-level zing of accomplishing something (*done it*) also grew into an aesthetic enjoyment or a sense of pride at creating order out of mental disorder (at feeling *on top of it*) in even one area of their lives. For these folks, admin pleasure came from having an effective approach to executive-function admin. Some of the items on their to-do lists might be admin (make that doctor appointment, fill out that form), while others might not (do laundry, take dog for a walk). But the keeping-track of all that needs to be done is a crucial form of admin that most every adult has to do —and that some people seemed to find some pleasure, or at least satisfaction, in doing well.[5]

These two forms of pleasure stem more from achieving the desired outcome than from the process of doing admin. This would fit our understanding of admin as valuable only for its results.

And yet some of my interviewees also suggested that the process of doing admin could itself be pleasurable, or at least satisfying. New father Joseph compared it to a crossword puzzle—observing that "it's not inherently impossible to like"; the problem is just "when it distracts you from the thing you want to do." One woman even said that she didn't see admin as *an obstacle* to doing things she cared about but instead experienced it as *a part* of doing things she cared about. She said this easily, without hesitation or qualification. I heard this kind of sentiment from a few other people, typically Super Doers, who seem to experience deeper pleasure from the process of getting on top of admin—or who are at least more in touch with that pleasure. Most of us aren't there (dare I add *yet*?). Can an inquiry into admin pleasures get us any closer?

To do that, we need to get more nuanced. The possible pleasures in admin vary widely. One concrete form of process pleasure recurred in my interviews, so we'll start there.

## Paper Loving

I was surprised by how many of my interviewees who are intentional about admin rely on paper. Paper calendars. Paper to-do lists. Little notebooks they carry around to keep track of things.

Pat, Vera, and Lillian are all in this camp, to name a few you've met. Interestingly, even Ellen, the Super Doer executive assistant to the Super Doer political couple, has gone "old-school" on her family's life admin—using a paper to-do list and a whiteboard calendar—to create a separation from the highly technological office space where she manages the political family's many calendars on her computer. Ellen enjoys doing the admin for her daughter because her world is "a simple fun world" compared to the "more difficult, more responsible world" of work and adulthood.

Will paper solve all of our problems? Not for me. And yet paper was part of solving my biggest admin failure.

## Light at the End of a Dark and Gloomy Admin Tunnel

I didn't begin this project looking for admin pleasures. As a Reluctant Doer, I hoped my research would lead me to find ways to get rid of some admin and do the rest more efficiently. But wrestling with my biggest admin failure brought me to a surprising truce with technology.

I have spent years trying to find a good system for keeping track of what I need to do. And for a long time, too long to admit, I had a truly horrible method. I would put urgent items into the calendar in my phone, recurring at a particular time, such as nine a.m. My aspiration was to complete the task, which seemed important, the next day. If I didn't do the item the next day, though, I would just keep it in that repeating calendar entry, adding more items as they came up. I tried to make a non-urgent list somewhere else, but I never looked at that list. So of necessity, I started to stick non-urgent items in the nine a.m. calendar entry as well. Every morning, my calendar would flash a reminder of the fifteen-plus items I was supposed to do, all mixed together despite different levels of priority. And whenever I did manage to finish any task, I couldn't easily delete it because my iPhone calendar wouldn't let me scroll earlier in the nine a.m. entry without deleting everything along the way—*since it's not made for to-do lists.* When the nine a.m. entry became too overwhelming and I needed to flag something genuinely urgent, I'd start a new timed entry—say, for eight a.m.

You can see where this is going. And why it was no good.

I felt inspired by my interviewees to try paper, but I wanted my running to-do list in my phone, not on paper, because I always have my phone. I don't want to waste time looking for the to-do list when I need to jot something down or have a minute to get something done. And yet the apps I have tried have exasperated me.

I tried an app called Any.do but it wouldn't sync with my laptop. Items appeared in one place but not another. I wasted two hours. An early to-do-list app I tried still sends me occasional messages reminding me to change the crib sheets for my child, who graduated to a big-kid bed years ago. The latest to-do-list-app fiasco involved Todoist, which sent me another email message today—to tell me I have thirty-two overdue items. The first of those items is eighty-one days overdue. The message begins "Hi EFE," as if we're friends.

Todoist turned out not to be so helpful largely because it drew me in with the prospect of many orderly-sounding categories, as if categories were going to make my mind or my life more organized. But it turns out that having many different headings to organize my to-do items does not help me. I don't sit down with a spare moment in the evening and say to myself: *I think I'd like to accomplish some medical admin tonight.* That just never happens.

Rather, I have an ongoing sense that many things need doing, and I either find or am given occasional windows of time in which to do them. Then I must quickly identify items of the right kind technologically or mentally; for example, business that must be done through phone calls is good only in daytime hours (though not on the subway), whereas something that can be done by email is fine anytime, including in a bout of four a.m. insomnia (but not if it's too mentally taxing for my tired brain or if it will keep me awake even longer). And yet thinking up those rubrics in advance is often too hard. If I can get the thing written down somewhere at the moment I think of it, I'm doing well.

Cleaning out the messy calendar-entry to-do lists and transferring all my to-do items to Todoist was itself a feat of admin-improvements admin. I started it during an Admin Study Hall, so I'd have moral support. People in an admin onslaught are least likely to have time and energy to make their admin more efficient. But I was persuaded by Danielle in a brainstorming session that process improvements can make *you* the future object of your own benevolence. That sounded pretty good. So imagine my disappointment when Todoist was a bomb, and one month later I was back to recurring calendar entries at nine a.m.

My process improvement eventually came from an unlikely source: a book called *The Artist's Way*, which was recommended to me by several people as an aid to writing.[6] After wading through some spiritual incantations, I hit upon an admin discovery.

A central technique of the book is something called "the artist's date." You carve out two hours each week to do something your "inner child" wants to do. The idea is to spur your creativity, especially if you're facing writer's block, by doing something playful, like seeing an afternoon movie or going running in a field with your eyes closed.

In the several weeks I managed to do these artist's dates—in an attempt to move my writing forward during a challenging time—my inner child pursued some conventionally playful activities, such as trying out paddleboard yoga and wandering through a new museum. Then one day my inner child wanted to do something truly perverse. My journal from the period describes it:

> Turns out that my inner child also cares about admin: My inner child wants to solve my to-do list problem. That little girl who liked stickers and back-to-school shopping wanted, last week, to go wandering through the notebook section of drug stores and stationery stores until I found an appealing notebook and some old-

> fashioned gold star stickers. So that's what I did. I went to three different drug stores, within a few blocks of each other on Broadway near my home, and settled on a nifty little notebook with bright flowers and a rubber band that keeps it closed.

On another occasion, my inner child wanted to go through my towering mail pile, deal with bills, and designate an antique wooden shovel someone gave me as an outbox for things to be mailed. That journal entry ended with the word *sigh.* A sigh of exhaustion. Also of pleasure.

It seems Ellen was onto something with going old-school on her parenting admin. Doing things with paper or on a whiteboard or designating an attractive physical space to put the mail—turning all this mental work into something physical—can make the adult sphere of executive-function admin feel a little bit playful, like a distant memory of something pleasurable.

Perhaps this surprise can help only those people who liked school supplies or who have a fondness for paper or wood. Then again, maybe there are more of us than we think.[7] Another Super Doer, Rosa, used to "try to create something pretty" out of admin—like a paper chart of goals for the week—and display it. Rosa's enjoyment of making her to-do list into a beautiful object is similar to the current trend of Bullet Journaling, with its Instagrammer groups of people who make multicolored artworks out of their organizing systems. That work looks pleasurable—and also very time consuming. One friend said my story made her think, suddenly, that her inner child liked color-coding. And if she'd just make time to color-code, she might manage to do that filing on her desk. Perhaps your inner child has some equivalent affinity. Perhaps it's elegant pens or a plastic in/out tray.

My conversion to paper was only partial. Before long, I realized that for the long-running to-do list, the Notes app in my phone works best. I always have it with me, but it demands no sorting or categories. It is, in a sense, very much like paper, but ever present and searchable. By contrast, paper works best for me on days with much pressing admin and actual windows in which to accomplish it. And paper adds an element of pleasure when an admin onslaught demands sustained attention or I have an unusual density of deadlines.

## Forms of Pleasure and the Pleasure of Forms

So should we all go and buy paper journals? Not necessarily.

Identifying our (typically elusive) admin pleasures involves recognizing our diverse tastes and ways. Here is an Admin Pleasure Inventory—a list of dimensions along which I've seen people vary in their possible pleasures (or displeasures) around the process of doing admin:

- ***High tech versus low tech:*** Do you prefer gadgets and devices to help with your admin, or do you prefer old-school tools like paper and pen?
- ***Beauty versus efficiency:*** Do you prefer to do admin beautifully or efficiently (when you can't have both)?
- ***Prompt versus thorough:*** Do you like to deal with something immediately, or wait till you can do it thoroughly? (One interviewee contrasted her husband's preference, which is to send "phatic placeholder emails" acknowledging or appreciating messages they both receive, with her own, which is to follow up later with a more substantial response.)
- ***Marathons versus short sprints:*** Do you prefer to tackle a massive pile of admin in one go, or do you prefer to limit admin-doing to brief bursts?
- ***Collaborative versus solo:*** Do you like the idea of doing admin with other people, collaborating or working side-by-side, or would you prefer everyone go it alone?
- ***Negotiating versus knowing:*** Do you like to haggle or ne-

gotiate over the prospects you face (financial or otherwise), or do you prefer to know the options clearly up front? Some work suggests that gender may predict responses to this question on average, but averages don't answer for individuals.[8]

- ***Managerial versus secretarial:*** When life admin needs doing, do you like making decisions, or do you prefer "administrivia"? One colleague told me his theory that all people prefer decision-making to task-doing, but, from what I've seen, tastes differ.

The invitation here is to ask yourself these questions and locate yourself on either side of each spectrum. Then experiment. See if there are ways to do admin more in the way you like. Perhaps you'll find new pleasures, or at least avoid some admin suffering.

Does the possibility of admin pleasures mean we should all just get in touch with our inner capacity for liking admin and invite in greater quantities? No. Not everyone has admin pleasures, and no one likes all admin—and even those with some admin affinities would often prefer doing something else.

Being able to find some pleasure in doing the dishes does not mean wanting to become a dishwasher. The grim part of grim satisfaction is that this work, even if it can yield some relief, typically doesn't feel like the *best use* of our finite time.

But recognizing admin pleasures can help us avoid the pitfalls of doing too much admin (when we don't have to) and hating it (when we do have to). Earlier we met the concept of pre-crastination—doing less important things too soon (or ever, I'd add). If we don't see the hidden allure of admin, if we don't recognize possible ambivalence, we may be more likely to get sucked in to pre-crastination. And if we don't recognize admin pleasures, we may miss a chance to make admin more tolerable when it's unavoidable.

*What's your admin-pleasure profile?* A surprising question, but one worth asking.

10

# Admin to Win Friends and Influence People

Alertness is the hidden discipline of familiarity.

—David Whyte, "Everything Is Waiting for You"

Enough about us. What can our thinking about admin do for other people?

The past few chapters have emphasized the benefits to the self—in times of distress or improvement or necessary admin-doing—of taking admin seriously. But thinking about admin can also help us to give and to persuade. Admin can be a way, recalling Dale Carnegie's classic advice manual, to win friends and influence people.

Asking the Admin Question—considering how admin might contribute to the problem or the solution in any situation—can improve *how* you give gifts and *what* gifts you give, as well as how you *receive* gifts. It can equip you with better ways to help people who are suffering, near and far. And it can help you get other people to do what they want—or what you want.

## Giving Gifts Better





I used to hate giving gifts to other people's children. Not easy to admit, but true. Before I had kids, I had no idea what kids of a particular age want. Once I had kids, I couldn't remember what kids of any other age want. In the past, I'd search online for generic ideas of good gifts by age; in a hurry, I'd just quickly order something inapt. Either way the task felt impersonal—just a big admin annoyance.

A few years ago, it occurred to me to give books as presents to kids. I love children's books, or at any rate the fraction of them I think are beautiful or funny or profound (or all three). Exquisite children's books are, in my sentimental view, the only poetry many adults read. Choosing these books for my friends and their children takes me through a mental process related to something I love. Now this task I used to dread has come to feel less admin-like and more like an end in itself.

Asking the Admin Question can also help you improve the gifting experience for the receiver.

Consider the admin involved in receiving flowers. Unless the flowers are already arranged in a vase—which can be expensive, and wasteful if the person has plenty of vases—the receiver has to think and organize: find scissors, identify a vase of the right height, and locate a good place for them (somewhere that they won't drip water on wood, say). If the recipient is in the middle of another activity, like greeting guests or cooking dinner, your gift may become a burden.

To turn that around, you might offer to set up the flowers as part of giving the flowers. Even if you don't know where to find the scissors and vase, you can "own" the task—taking instructions until you complete it.

Same goes for gifts for other people's children. If you bring a toy, offer to open it and figure out how it works. Provide or find batteries if needed. And if you give someone technology, be prepared to get it to a functional level. In other words, think about the admin entailed in receiving and making operational a gift that you're giving.

Some hosts won't want you to set up their new cordless speaker or rifle through their cupboards looking for a vase or scissors; they'd rather do it themselves. But for other hosts, amid whatever else needs doing, nothing would be nicer than to escape the added burden of showing appreciation for a gift by figuring out how to display it or make it work.

## Giving Better Gifts

If you're up to the task, you can use admin to ratchet up the level of generosity. You can give the gift of admin. Ava, a super-dynamic young woman I interviewed, described two of the best gifts she ever received. One: A friend created a webpage for her. The friend knew her well enough to pull this off, crafting something true to Ava's taste. Two: Her mother noticed her using a smashed iPhone and insisted on taking it to the Apple store to get a new screen. Ava, who was managing a dizzying array of commitments and didn't know when she would get to these projects, was overcome with gratitude.

Rather than doing a particular admin task—like creating a webpage or arranging repair of a phone—you could offer your time for admin-doing of whatever sort. Did anyone else give "chore coupons" as a kid? For Mother's Day, I would cut coupons from colored construction paper and mark their currency like real gift certificates: *Good for one free dishwashing. Good for fifteen minutes of sweeping.*

How about giving a loved one an admin coupon? *This entitles the recipient to one hour of admin-doing per your instructions.*

If you know the person well—which surely you must in order to give something this corny and generous—then you may also know if the person would prefer shifting (where you take over the task), sharing (where you join forces and do the person's task together), or supervising (where you just provide company or accountability, remotely or in person, while the other person does her task).

Remember your admin personality, though. If you aren't going to do it, don't offer. When Zack was asked in his interview if he did admin for anyone else, he thought for a while, asked for more information about the question, then laughed and said, "Friends don't ask friends to do admin."

His wife, Livia, in a separate interview, immediately rattled off the names of friends and family whose medical admin she had helped with. For a close friend with cancer, she had researched which doctors to go to, called the doctors, gone to appointments, and researched treatment options. She noted how difficult, and painful, it could be to be your own medical advocate while going through wrenching health-care challenges.

Even if you're up to the task of offering admin, take a lesson from the Avoiders and Deniers. Imagine how much time you'd like to give—say an hour—then consider cutting that in half. And half again. With any luck, fifteen minutes will be enough time to do some research on cheaper internet services—or whatever small task the recipient has been putting off. If the first coupon goes over well, maybe next time you'll offer more.

## Giving Thanks Better

Making admin visible could also change the way we say thank you. Empirical psychology and Buddhist teachings—among other traditions—tell us that practices of gratitude make us happier.[1] It's a deep question, though, how to connect the production of thank-you messages of one kind or another with practices of gratitude that actually serve givers or receivers.

For starters, formulaic or insincere thank-you notes might become outmoded if we all begin to see them as the pure admin they often are. Not every gift is appreciated. Some are even thinly veiled criticism. (Think ear-hair clippers or slimming undergarments.) Many more gifts are simply thoughtless or routine gestures or essentially business transactions. Trying to turn these transfers into moments of deep connection with an unsuitably heartfelt reply seems bizarre.

Pro forma thank-you notes could be replaced by a wide range of ways to show gratitude—including, for some, by agreements to express the closeness of a relationship by deciding to skip the thank-you notes. (My mom and I did this recently.) Both would know that their closeness enabled this arrangement. Saying thanks for thoughtful gifts would not require vast time or energy. For instance, sending a photo of a gift, once opened, lets the giver know you received it—and sooner than a handwritten note—and could even show the person how you're using it (or wearing it) or perhaps give back a little beauty (this works well for flowers).

And for kids? What's the best way to raise them into a sincere and admin-conscious culture of gifting and appreciating? I'm still puzzling over this, but these basic ideas seem promising: The tenor of the gift should be reflected in the response, and the experience of the thank you—for receiver and giver both—should matter more than formalities. Writing notes, or at least signing them, might be meaningful for a child just learning to write his name; choreographing a thank-you video might be fun and playful and even artistic, depending on the kid. Simpler still is sending photos or short videos of children opening gifts. If you opt for videos rather than pictures, though, remember that videos take time to watch—delightful for some but admin for others.

## Better Serving the Bereaved

Lillian had been to a funeral the week of our interview. She found herself reflecting on death admin.

She first praised the amazing people who helped the mourners —by cooking meals, for instance—and also those who did the admin to coordinate the helpers. Then she began to think about ways an admin perspective could have allowed friends to better help the family through this difficult time.

What if, she mused, someone put together a list of all the invisible admin that grieving folks are dealing with and the ways you could help with what was usually a painfully isolated activity? We imagined together people willing to assist—by shifting, sharing, or supervising—the widow with the pile of mail in her hallway. Volunteers could be organized to deal with the bills and banks. Or to puzzle through the life-insurance admin. Or to help figure out what to do with the deceased person's minor possessions—the ones that weren't valuable enough to bequeath but feel too personal to throw away.

Having admin done well for you can be a powerful experience. Several interviewees recounted having this feeling, even when money changes hands. Barbara is still close to someone who did admin beautifully for her at work, a person who made her feel cared for and like she had a really strong support system. Lillian and I were imagining how that kind of caring through admin might feel, in a time of need, from an already loved friend or family member.

Death admin is a striking example of how one person's approach to admin during life—how much a person thinks about his own death and plans for it—can have dramatic effects on others. Whether a person has a will, how detailed it is, whether he has managed his money well, how much guidance he's provided for his funeral—these choices have a profound impact on whoever winds up with the admin following his death. Professional services have been created for people who want to plan their own funerals. These are often profiled in the media as solipsistic curiosities. And yet this seemingly morbid admin may greatly relieve the family members or friends who would otherwise have to guess what their recently deceased loved one would have wanted.

## Getting People to Do What *You* Want

Admin can also help you sway others in your direction.

When I was thirteen, I told my father I wanted to change schools. He asked me a few questions, enough to understand that my reasons were pretty superficial. Then he replied, "Okay, you get in touch with the guidance counselors at the new school, learn about the honors program, find out who the best teachers are and what courses you would want to take. Then come back to me and we'll talk more about it."

We never discussed it again.

My father set up admin barriers to my plan and I abandoned it. I understood on some level that he was trying to make it hard, but his requests were not unreasonable. They were things that he and my mother would have tried to do for me, or with me, were I to move to this new school. By asking the Admin Question thoughtfully, he was giving himself a diagnostic tool to measure my seriousness, but he wasn't merely throwing trivial admin drag (sludge) in my way. He was identifying the work involved in thinking this through concretely and putting it on me.

Would he have let me switch schools, had I gotten that far—in other words, was his plan cynical or heartfelt? To this day, I don't know.

Imposing admin drag can deter others' efforts. Does anyone else remember the old-fashioned Record of the Month Clubs, which gave you an amazing deal on some initial records (maybe even a penny per record) on the condition that you had to buy or ship back future monthly overpriced offerings for a year or more? In other words, you had to make a monthly trip to the post office or pay big on the back end. Those companies understood the role of admin drag in influencing people.

I am not advocating that you do this with your friends. But maybe, ever so carefully, you can do it with your children.

## Getting People to Do What *They* Want

You can also use the Admin Question to help figure out what you want—or to help someone else figure out what they want. In this spirit, I've asked it of male friends who want their wives to find jobs outside the home again after kids. Yes, I have known more than one of these guys. I'm sure there are men who wished their wives didn't work outside the home and some who are glad their wives don't. But there are also men who think their wives would be happier returning to an endeavor they loved before the kids came along or who miss having conversations about topics other than the kids and the home. (Or they may desire the extra family income; it's complicated.)

When I hear my friends say this, I ask, "Have you offered to help find the childcare?"

I encourage these guys to help take ownership of a childcare search—the stress, the strain, the ambivalence. The incredible feeling of responsibility. Anyone who has looked for childcare knows that even if you are fortunate enough to have money to afford great help, these searches can be a logistical and emotional nightmare. For someone with mixed feelings about leaving the kids, that search may be an insurmountable barrier. Zack, who has a two-year-old, nominated childcare searching as the worst kind of admin: "It's also doubly stressful because you have the sort of annoyance of the admin, and you have the emotional issue with your child—you have the emotional issue with 'Am I a good parent?'"

So if someone in your life needs some or more childcare—whether it's your spouse or your friend or your adult child, and whether it's because you think the person should go back to work or school, take on other projects, or engage in some self-care—consider asking how you can help with the practicalities. Be prepared to help draft an ad, find places to post it, screen online responses, search for childcare centers, visit centers, assist with interviews or evaluations of centers, and talk over the best options.

If you take earnest steps in these directions, and your target person still doesn't close the deal, then maybe you've learned something about that person's wants. And if you can't get yourself to do any of that admin, then maybe you don't really want the person to make the change that badly. At which point, I gently suggest you give the person a break and offer some appreciation instead.

## Making Change Beyond the Individual

Asking the Admin Question can enable more than a gift to one other person. Paying attention to the role of admin could lead you to creative ideas and broader impact in whatever arena matters to you.

For example, a college student, Lenzi Sheible, created an organization to help overcome the admin barriers to abortion for women living in parts of Texas where clinics were closing in response to new legislation. As part of funding their travel to far-off clinics that remained open, Sheible also did their travel admin.[2] As another example, in some jurisdictions, adoption advocates have worked to make adoption easier for the adopting parent or the birth mother or both, since the paperwork and procedures for adoption can range from perplexing to overwhelming.[3] And advocates who gather and disseminate information about the treatment of workers or animals involved in food production save concerned consumers the admin of researching their food choices.

More generally, taking admin seriously can help increase support for any cause. Empirical research suggests that asking the Admin Question and reducing minor impediments to charitable giving (for instance, admin hassles) can substantially increase donations.[4]

## Getting Better Service

Back to your admin. When you're trying to influence someone—whether it's an airline gate agent who wants to bump you from a flight or a landlord who wants to raise your rent—what's the best tack to take? Is it to throw your weight around, show a little swagger, flash your business card, tell them you're a lawyer/cop/boxer? Does yelling help? Crying?

Many people have told me their "speak to the manager" stories. The legal advocates in a brainstorming session I ran offered competing versions. One man talked about needing to push past the "idiot robot-stamper." By contrast, one woman emphasized being "syrupy sweet and pretending you don't know what you're doing"; the desk clerks have been "yelled at so many times . . . sometimes doing the exact opposite can help." The rest of the group wondered if gender played a role in what strategies worked for each of them, which certainly seems possible in light of the studies finding differences in how men and women are perceived when they negotiate—and depending on *how* they negotiate.[5]

Some research supports a humane approach in retail-service encounters, since emotions are "contagious." This work suggests you'll get better results by adopting an *affiliative style* (engaging in a "warm, friendly manner") than a *dominant style* ("trying to control or dominate").[6] I interviewed one woman, Sophia, a subsidized-housing advocate who seemed to have remarkable influence with an especially tough crowd: housing bureaucrats and landlords. She told me about marching straight up to the eighth floor of the housing office to report on a frontline clerk who demeaned her client. Her apparent warmth was consistent with the research supporting an *affiliative* approach, but our interview convinced me that the most powerful version of that approach is not being friendly in a shallow or superficial way. Her manner combined a respect for all people with a dignity, and a deep faith in her convictions, that gave her interpersonal authority. You knew, just talking with her, she was a force to be reckoned with. I don't expect to develop gravitas like that, but, since interviewing Sophia, I have gotten better at approaching admin interactions with two values in mind: on the one hand, a clarity and conviction about what I want, and, on the other, an appreciation of the other person's humanity, a basic sense that we are in this together.

Even if a clerk doesn't have much authority, you may make progress by asking for his or her name and saying, "Are you the person I can follow up with?" These questions can help clerks feel they have more decision-making power and accountability. Moreover, taking a name can give you leverage where you might otherwise lack it later on. I've heard many stories of people holding retailers or even medical providers to a price quoted on the phone just because they wrote down the name of the person they spoke to. It happened to me recently with a doctor whose office said he would accept my insurance, and then, on leaving, I was handed a bill for $700 and told to pay on the spot. Fortunately, I had made a note of the name of the person who'd helped me, and eventually I was cleared.

Stories of admin hell often end with reaching "the Person," as Melinda, a new mother, put it when she told her story of six months of calls to follow up for an insurance nightmare with a miscoded diagnosis. After she finally asked to speak to a supervisor, she got to the Person. "He was great. And then he was like, 'You won't get any more bills about it, or calls, until it's resubmitted.' And I didn't." It all went smoothly from there.

Finding the humans in any admin situation, and treating them like humans—humans with a name—can go a long way toward serving them and serving you.





## Rethinking the Admin Question

At the end of Part 1, I noted that one takeaway of this book might be that you should avoid admin. Don't answer your phone or your mail or your email, and maybe you'll get away with doing (even) less.

This might be a good idea in particular realms or particular lives. Some admin literally goes away when you avoid it. Emails inviting you to join a social club eventually cease if you don't respond or if you send a quick *No thanks.* Other forms of admin just fall to someone else.

Before clearing the path to total avoidance, though, bear in mind the stories you've heard about building friendships and influence through admin. And as we transition to a chapter on love and relationships, consider some advice offered by my interviewee Livia.

She and her husband, Zack, struggle over admin—her doing it, his not doing it as much. What would she say to Zack about admin if she were his friend rather than his wife?

"I would tell him: Do more because it will make Livia happier, love you more, want to have sex more."[7]

Part III

# admin futures

11

# Relationship Tips

It is in these acts called trivialities that the seeds of joy are for ever wasted.
—George Eliot, *Middlemarch*

You may not want relationship advice from someone going through a divorce. It's practically disqualifying. But I've learned a few things while working on this book, mostly from the many generous people who let me interview them and the many more who informally invited me into their lives and relationships. Trust them if not me.

It is, I have come to think, worth asking the Admin Question when choosing a mate. In other words, it is worth asking, *What role does admin play, or could it play, in the problems and strengths of this particular relationship?* It is also wise to think carefully about admin at foundational moments, like moving in together or getting married. And if you're already in it, till death do you part, there are admin experiments you can try when you feel inspired—or desperate.

You may read this chapter with romantic relationships in mind. But admin affects our relationships across all contexts—friendships, families of origin, housemates, coworkers. Many of the ideas here apply equally to relationships that never had any romantic element.

I cannot promise peace, love, and harmony. I do hope, though, that if you ask the Admin Question in your relationships, take the answers seriously, and open up to experimentation, you might just avoid some mistakes that others have made before you.

## Admin in Love

Many couples I interviewed struggled with their interactions around admin. For some, admin was a major source of friction. For my ex and me, that was certainly the case. Indeed, I had begun writing this book to solve the admin problem in my marriage. The fact that my marriage ended might suggest that admin problems were the source of our problems.

Or, worse yet, as I have sometimes worried, perhaps my attempts to *solve* our admin problems were the *source* of our problems. This was not merely the rabid delusion of a suffering mind. A colleague of mine had actually predicted such an outcome years earlier. On hearing about this project and my attempts to make admin visible in my marriage and life, he had proclaimed, in the public forum of a faculty workshop, "That sounds like a one-way ticket to divorce!"

Hideous as it was to remember the colleague's prediction once my divorce was under way, it didn't take long before I realized he wasn't quite right. Admin didn't cause my marriage to fail. Nor did trying to solve our admin problem. Admin is important, but as a means to an end; it could hardly, by itself, be powerful enough to wreck a marriage. And yet admin was not irrelevant to the story. Our admin problems were reflective of the fact that our marriage wasn't working.

This is the simplest truth about admin and relationships. Admin dynamics reflect other dynamics in a relationship.

In my interviews, I met some couples who seemed genuinely happy in their admin dynamics. This was jarring. I had encountered so many people for whom admin was a cause of tension or resentment. What made the happy admin couples happy?



One couple seemed brilliantly matched in their admin personalities. The politician Nate and his wife, Cheryl, the Super Doer couple we met earlier, both invest in their separate and joint projects an unparalleled amount of energy. Their shared approach to planning and scheduling and prioritizing in life seems part of what brought them together. But Nate and Cheryl were kind of hard to relate to, even if they were fun to admire from afar.

More accessible was a couple like Josh and Daphne. They had an ease about them, a comfort with each other, a curiosity about the topic of admin and where our interview might lead them. I interviewed them together, and, although I can't know for sure, I doubt their answers would have been much different if I had interviewed them separately. The only change I'd expect concerns praise: They might have been a little less complimentary of each other. They seemed very aware of the impact their words had, both joking words and words of praise. They seemed very aware of each other in general.

Trust was the strongest feeling I got from them. Then caring, warmth. They had been married nine years, and they had three young children. The concept of admin resonated for them both. When I gave a few examples and asked what admin their lives involved, she said, "All those things; all those things you just said."

For the most part, Daphne directs things at home. Josh works long hours outside the home, and she doesn't. Generally she is the list-maker and he is the task-doer for his share of the household admin and chores. (He does manage long-term financial matters, like investments and planning, while she deals with routine financial matters on top of day-to-day spending.) She keeps a running to-do list on the fridge with a little section labeled "Josh." He sometimes strays and starts a household project that isn't on the list, and she reins him back in. In our interview, they laughed about this. They laughed a lot, in fact, including about how Josh *does* multitask sometimes—because he does his admin in the bathroom. Neither of them thought the other is extraordinary at admin. Both seemed to trust the other to get a job done, though, and to do it well enough.

They each appreciate what the other does. When I asked Josh what it felt like to have someone do admin for him, he paused meaningfully, seeming to breathe it in, and then said, "It feels like someone is taking care of me when someone does my administrative work." Daphne described how well she juggles some complicated aspects of their lives, like the morning routine. She seemed proud, sweetly so, like she knew Josh hadn't fallen for her for her organizational skills, but still, she'd gotten pretty good at managing things. And she liked that he appreciated it.

Josh and Daphne's story is not a story just about admin. Theirs is a story about loving and respecting and trusting each other in ways that translate into their relationship around admin.

So is that it? Find a good relationship and you'll necessarily have a good relationship around admin? No, not from what I've seen. If you aren't careful, admin can become bad even if the relationship is good. And admin can be the worst place of friction in a relationship that has a mix of good and bad—that is, almost every relationship.

In my admin travels, I've been particularly struck by people who thought and even talked before marriage about who would do what in their household, including tasks that fall into my category of admin. I think, for instance, of Erin, who asked her then-fiancé Paul to take on a job she really didn't want: taxes. Erin and Paul have also found ways to share admin projects that seem to free her from the kind of admin frustrations I've heard from many others. For summer camps for their children, whose different ages require building several unique schedules, Paul gathers all the catalogs as they come in and creates a giant spreadsheet of the possibilities for each child, then they sit together and make the decisions, and then she enrolls them and fills out the forms by the respective deadlines.

Erin clearly does a lot of admin, like managing the household calendar, but she spoke with a lightness about her husband's calendar questions. "What's the schedule for tomorrow?" he'll say, every night, by Erin's account. "Look on the calendar on the computer," she'll say. "What's the schedule for tomorrow?" he'll repeat. And so on, until finally "I tell him the schedule." This might sound exasperating, and perhaps it is, but Erin laughed easily as she told it.

I also think of Vera, working to understand how her fiancé Saul's acts of admin service are his way of saying he loves her—so she can see them now and appreciate his efforts. This recognition seems, in turn, to inspire her on her journey to becoming more of a Doer for love.

So what can anyone do to have an admin-happy relationship? For those of you who haven't met your match yet, and who want one, let's start at the beginning.

## Choosing a Mate, or Popping the Admin Question

Few people are going to choose a romantic partner based on the prospect's admin-doing or non-doing. In defense of anyone who does so, though, it's worth noting that people have chosen partners on far more suspect grounds.

1. **Consider asking the Admin Question when finding a mate.**

   Admin as a factor in courtship sounds pretty unromantic. And yet choosing a life partner is not just about deciding who you will have sex with for the rest of your life (if you're among the fortunate ones who make it that far in marriage, and in sex). Choosing a life partner is also joining a team.

   Research suggests that your partner's *conscientiousness*—

a dimension of personality associated with "being careful, thorough, responsible, organized, and planful"[1]—predicts your job satisfaction, income, and likelihood of promotion.[2] Some work has found that a partner's degree of conscientiousness predicts relationship satisfaction as well.[3] (The one study I know that finds no connection between a partner's conscientiousness and relationship satisfaction studied college students—people who likely don't share that much admin yet.[4]) A recent study also finds that one's own conscientiousness predicts sexual function in both men and women—and a steady *partner's* conscientiousness predicts sexual function in women.[5]

Let's pause and think that over for a moment. Your *partner's* personality, along a dimension that seems highly relevant to admin, shapes *your* future prospects for success and happiness at work—and possibly for happiness at home, too. Perhaps this shouldn't surprise us. You're joining a team that involves running a life and a household. Everyone's contributions matter for everyone else. And as that young father warned us in Chapter 4, *Admin* is *marriage.* (This idea of marriage as joining an admin team probably informs Nora Ephron's infamous line "Never marry a man you wouldn't want to be divorced from."[6])

For sure, if you're merely choosing whom to bed once, then a prospect's admin personality may not matter much to you. Though be warned: If you are instead thinking of becoming sexually involved in an avowedly polyamorous relationship, you should be extra careful about your partners' admin abilities. If you're an admin Doer, and your poly partners are not, that is a lot of extra admin for you; just ask Alice, the woman in my brainstorming session with an Avoider husband and a Denier live-in girlfriend. Moreover, the scheduling and emotional maintenance of a polyamorous relationship itself involves a lot of admin. A poly woman I interviewed, Bonnie, worked as a legal secretary, and her husband, James, is a project manager. These are serious Super Doers.[7]

Certain sections of this book are haunted by a particular person. As I write about choosing a mate, I see Rosa, the Super Doer oldest child who grew up translating for her immigrant parents. Rosa, now studying for a postgraduate degree, thinks partners should divide up work based on who has the time for it. She gets upset at her current boyfriend: "I don't understand how someone can go about their laz—their days without planning ahead." (Yes, it did sound like she slipped and started to say *lazy* instead of *days.*) It sounds like Rosa might be dating an Avoider. But she hopes that someday, when she and her partner both work jobs with long hours and her partner sometimes gets home first, those nights he'll be the one "who creates that grocery list, or goes out and takes care of items, figures out whether we need to send anything to the dry cleaners." I hope that Rosa can create a life that supports her dreams, at work and at home. I hope she asks the Admin Question before settling down.

2. **Ask prospective mates how they do admin and feel about admin.**

*When do you do admin? How do you remember to do things? When do you go through your mail? Do you know anyone who does too much admin? Too little? Has your relationship to admin changed over time? Do you expect it to change in the future?*

These questions may sound innocuous, even mundane. But Admin Questions may be the stealth equivalent of "Tell me about your mother." Admin touches intimate places, and most people rarely talk about it. Raising the topic may bring you closer. Or at least teach you something. (And if you're already partnered, you might consider supporting your single friends in having these conversations by asking if the people they're dating are good at life admin.)

3. **To predict how someone will do life admin in the future, look at how they currently do life admin, not how they do their *work* admin.**

I've encountered plenty of people who are competent at work admin but are Avoiders or Deniers at home. Partners often lament this quality in their better halves—like my colleague whose husband manages a huge staff at work but can't manage a playdate.

Seth, a married father of two, talked about his evolving appreciation of the admin part of his job as a consultant—the process and people-management side. He has come to see in recent years how integral process is to outcomes. And yet, he reflected, "I just don't feel the urgency to be lights out on admin at home." (For those who don't follow sports, *lights out* in this context means "giving a great performance.") Seth asked, rhetorically, "Would it be worth the effort?"

Two women in a brainstorming session in Glasgow realized they were opposites in where they do admin best. Fiona said, "I feel like I can only be efficient and competent at work by ignoring all the life admin that you're talking about until it becomes so desperately urgent that it has to be done." Hearing Fiona say this, Bridget worried that she was "slightly the reverse." Bridget found life admin appealing for "the real sense that you can complete the task," in contrast to the long-term projects of her professional work and, of course, parenting. (Young children, another participant chimed in, "never allow themselves to be neatly ticked off" the to-do list.) If you met them at work, you might think that Fiona's a Super Doer and Bridget's an Avoider, but their personalities at home were the reverse.

Work admin habits aren't a reliable predictor, so look instead to how someone does his life admin. Someone who doesn't do admin in his own life probably won't do it in yours. Yes, people can change. Vera is becoming a Doer for love. But not everyone on an admin trajectory moves toward more admin-doing, as we saw with Lillian, who became an Avoider after her divorce. So Non-Doers may become Doers, but I wouldn't count on it.

4. **Someone who has strong views may create extra admin for you unless he is prepared to do the associated admin himself.**

Kurt, who owns a business, suggested that his strong views about household matters created some of the conflict in his marriage. His view on grocery delivery: "Ew." He thinks it's "gross food" and believes in buying your groceries at the store. His view on screen time: Kids should be given stimulating activities—learning an instrument, playing sports—rather than being allowed to "sit at home and look at a screen all day."

Kurt also suggested that admin-doing was contrary to the philosophy of his family of origin. "Winging it is part of my family's MO," he observed, or at least it was for his brother, his father, and him. "And all of these people who schedule stuff absolutely sort of kill it for us." He evinced self-awareness about his ambivalence, musing that not having a plan "is a thrill but also some of it is maybe not wanting to do that work." Admin was a "bone of contention" in his marriage, which ended in divorce.

5. **If you don't trust how someone does things, then it will probably be hard to trust that person to do your shared admin.**

Getting someone else to do admin to your standards or in

the way you prefer is awfully difficult. Admin is hard to see and often hard to do together. Trust becomes especially important.

In one moving interview, new father Zack described his fantasy that admin could be romantic. He hoped it could be, anyway. He seemed grateful that his wife, Livia, could be trusted with their household admin, since none of his previous girlfriends could have been.

You might think Zack chose his wife for this reason. But he seemed almost surprised by his good fortune in ending up with someone so competent in this domain that he had come to realize carried such significance, now that they have to find affordable childcare and juggle two jobs with complicated schedules. "I think you have to trust somebody a lot. You're allowing them to make a judgment call for your life."

So what would I say to the hopeful Super Doer Rosa, who aspires to equality and is dating a guy who frustrates her because he just doesn't plan? If you decide this is the guy for you, go in with your eyes wide open. For starters, pay close attention to foundational moments.

## Foundational Moments

Moving in together. Getting married. Having a child. Big events like these help to define your relationship. They also set patterns around admin—patterns that may stick. And they offer opportunities to make choices about those patterns.

Depending on the relationship, your foundational admin event might be something less conventional—you might host a party, plan a trip, or get a pet together. And if you are on the proverbial path from love to marriage to baby carriage, your way of doing each event might be more or less admin-intensive.

1. **Consider the admin quantity and quality of your options.**

*Destination weddings.* Prognosis: Admin-heavy. That admin might feel like planning an exciting trip together. Or it might feel like trying to apply for a driver's license in a country you don't know from someplace far away.

*Weddings in one partner's hometown.* Prognosis: Generally admin-heavy for the hometown partner and probably for that person's family. Travel-admin heavy (and potentially expensive) for the out-of-town family.

Chris and Melinda started to plan a wedding, then ditched the prospect of even a small wedding, opting for city hall instead, largely on admin grounds. Eleanor said she abandoned her preferred marital name-change plan because the admin was just too much.[8] Baby-making can be more or less admin-intensive as well, though often not by choice. One new father through surrogacy described their multiple rounds of in vitro vividly: "At the time it wasn't clear when, if ever, our lives would be free from that sea of admin we were swimming in."

What kind of process or event you undertake at key moments will of course depend on your particular values, hopes, and options. But asking the Admin Question may affect the path you choose—and at least it'll help you know what you're getting into.

2. **Keep in mind that admin is sticky.**

Recall Rena and Kevin, the couple whose cross-country move—taking turns driving and setting up utilities for their new home—still determines their division of labor for bill paying. Or think of Lauren, whose housemate's attempt to do landlord admin for Project Pigeon "boomeranged" back to her because she had lived in the apartment the longest.

Many couples with children report that those early weeks or months or years when one parent was at home set patterns for admin as well as direct childcare. The parent at home became the expert on what was needed and got used to giving or obtaining it. Some other parents use admin's stickiness—plus the flexibility as to when and where it can (often) be done—as an opportunity to go against the grain.

Admin is sticky. Bear that in mind when you first move in together, and when making that first call to the plumber or the vet. Think about stickiness when you start a joint grocery list. Should it go in a notebook or on a whiteboard on the fridge? Or in a list-making app? Even if the app can be shared across two or more people's phones, will the users all be equally adept with the technology? This brings us to skills.

3. **Gaining skills can make *you* sticky.**

Admin is information-heavy, whereas chores are often skill-based. That is why you can do dishes in a stranger's home with greater ease than you can make someone else's grocery list. But admin also involves skills. Having certain admin skills can make you the initial Doer or the emergency Doer, and then that job may stick to you.

Multiple interviewees—women—told me that they do X or Y admin in their homes because they are "better with computers." These women have husbands who work in offices and use computers there. But something about perceived differences in technological agility helped to create patterns of admin-doing that fell on one partner.

One lesson: Be mindful of what skills you have and which ones you decide to teach the other person. If you're the only one who knows how to use the scanner, consider giving a lesson in scanner use. Even if you decide to keep scanning, the reasons for your admin-doing aren't obscured by any sense that you are the only one who knows how. The thank-yous might become more generous.

4. **Be especially thoughtful about your choices around any outsourcing, paid or unpaid.**

Managing outsourcing typically falls on women even more than the doing of the underlying activities, as discussed in Chapter 4. Men partnered with women are less likely to manage a housecleaner (or a babysitter) than to clean house (or care for kids). And women are more likely to be involved in household repairs when the task is to manage outside help than to do the actual repairs.

Whatever the gender makeup of your relationship, pay attention to who makes those calls and who builds those relationships. Who assumes responsibility for collecting and communicating information for whatever needs doing? Whose contact information is given?

Consider Dorothy. More than twenty years ago, she and her husband disagreed about whether to clean their own apartment or outsource the task to a professional. He refused to do the cleaning, saying, "I've got better things to do with my time." (He went on to found an important nonprofit organization, she said, so his words were not mere puffery.) In the end, they hired someone. Ever since, every two weeks, Dorothy prepares breakfast and lunch for the cleaner, which involves a special grocery trip, because the cleaner eats things they wouldn't have in the fridge otherwise; and Dorothy spends time listening to the cleaner's problems, of which there are many. In retrospect, if she could do it all over, Dorothy isn't sure she would set this pattern, but she's not prepared to change it now.

The lesson here is not *Avoid this kind of interaction.* Some people want it or feel it is important, or the right thing to do. (This is even more likely if you outsource household labor to friends or relatives.) Other people like to keep these interactions brief and professional. The lesson here is to choose carefully, knowing that admin is sticky and that many relational patterns around outsourcing follow gendered patterns.

Last, if your household does outsource, notice whether you ever say thank you to the partner who manages that relationship, if it's not you. Unless you ask, it may be especially hard to see the time and mental labor involved.

5. **Ask yourself if you or your partner might want to talk about any of the divvying up.**

   Research suggests that relationship satisfaction is predicted by whether or not you had conversations about household distribution of labor *that you wanted to have* before marriage.[9] This is not surprising. Yet many people don't have these conversations.

6. **Look for the fun. And if you can't find it, divvy up the admin of making it so.**

   How do you make admin visible and not suck the joy out of this foundational moment? Can you find the fun in the admin, or if not, can you make your own fun, as they say? My vote: Turn on music, serve food. Then talk.

   Another idea: Gamify it. Last one to finish the tasks buys dinner for the next planning meeting. First one to issue a (pestering) reminder before an agreed deadline brings flowers next time. Whoever delivers the most criticism or the least appreciation has to find an appealing venue. Play with it.

## Once You're in It

Okay, you're already deep into a relationship. Maybe you had the admin conversation at foundational moments, maybe you didn't. Are you stuck with your patterns? Not necessarily.

1. **Relativize your position as an idiot or a maniac.**

   You and your partner may well have assumed some roles —Doer and Non-Doer, maniac and idiot—in general or in some areas. If you are not the Doer in one or more areas of your life, notice how fun it can be to be the Non-Doer. Not just because you get out of all that work, but also because you get to feel like the cool, nonchalant person rather than the anxious person up at 4:00 a.m. researching what could go wrong or planning how to make things go right.

   Also notice how appealing it can be to be the Doer. How smugly competent you can feel if you're the only one who knows what's going on and when and how—how tempting it can be to feel you're the only adult in the relationship (and perhaps a long-suffering one at that).

   Noticing one's relative position may not change things for everyone, especially the strong Super Doers and Deniers. But many people along the admin-personality spectrum can gain empathy and insight by remembering their alternative locations across time or contexts or relationships.

2. **Play the Listening Game.**

   Everyone says listen. Listen to your family; listen at work. Listen, listen, listen. But how? And what if you have no time or no patience? Some listening advice is very demanding. Erich Fromm tells us in *The Art of Listening* that the listener "must be endowed with a capacity for empathy with another person and strong enough to feel the experience of the other as if it were his own."[10] If you can do this, good for you. But if listening deeply sounds beyond your reach, good news lies ahead.

   There's a simple exercise you can do. It's time-limited. It can be adapted to many purposes. It can be turned into a game. Or a conversation starter on date night.

Here's what you do. Set a timer for three minutes. The first person talks, uninterrupted. Then the second person reflects back what was said for two minutes. Then reverse roles. No agreeing, no disagreeing. When you finish reflecting back, you can ask, "How did I do? Did I miss anything?" Ten minutes, total. If you love it, if you both have more to say, you can repeat.

The topic of this exercise can be anything. When my adult nephew and I played this game while waiting for his sister to cross the graduation stage on a steamy day in May, our topic was air conditioning. For admin topics, you can ask general questions—*What admin's taking up your time right now?*—or pragmatic ones—*What should we do about your ailing father's living situation?*

The timer can feel artificial, but the timer is your friend. It lets you know this activity can't take more than the minutes you allot it. And let's face it: For many of us, listening is hard. The timer lets you know you have to listen for only X minutes longer before you get to say your piece.

What's the point of this exercise? Playing the listening game with admin can make admin visible. You can feel seen for your admin-doing. Or seen for your reasons to resist or resent admin-doing. But the structure can also keep you from getting bogged down in a debate. In the words of a wise person I know, *it means more to be heard than to be agreed with.*

One study suggests that when couples talk about their household distribution of labor, women married to men may *feel* more satisfied.[11] Talking, without changing the actual distribution much if at all.[12] Some might find this result disturbing; others encouraging. Whatever your view, it's worth bearing in mind.

3. **Look for simple ways to transfer information seamlessly.**

Information transfer is a big part of admin's stickiness. If you can both access the same information, then admin becomes more visible and potentially less sticky.

Some interviewees gushed about their information-sharing technologies. More than one couple who had implemented a shared calendar couldn't remember how they had managed before. It seemed archaic for them not to both know what was happening without asking each other. If you're both tech-savvy (and tech-happy), Wunderlist, Dropbox, and Google Calendar (or their latest equivalents) may be your friends. If not, old-school methods may be a better bet for getting everyone's buy-in: a grocery or to-do list taped to the fridge; a wall calendar with dates. (Someone who wants a portable version of what's on paper can take a photo of it.)

Whatever method you use, making information mutually available is a steppingstone to experimenting with unsticking tasks. Even if uptake is imperfect, at least both people have the option of saying, "Look at the calendar," and sticking with it.

4. **Experiment with planning meetings.**

Getting together to do your shared admin may sound unromantic. But one advantage to designated planning meetings is that they can cabin the admin. Someone can say, "The committee is not in session" at other times, like on date night.

For the introverts among us, though, getting together to do admin is going to sound inefficient, unappealing, or worse. The next section is for you.

5. **Play with negotiating.**

If you want to divvy up tasks or domains rather than doing things together, you need to find a way to do that. One option is to negotiate. Try making it playful. Asking, "What's it worth to you?" can open the door to collaboration. New possibilities may emerge for (what negotiators call) *enlarging the pie.*

And sometimes a swap can help to unstick you. Bella told her brainstorming session about getting her husband to trade off months doing their household budgeting—something she'd been wanting for a while—by agreeing to take out the garbage (usually his job) for two months. Sometimes everyone can benefit, rather than the outcome being entirely zero-sum.

6. **Try shifting who drops the ball. And whom the ball falls on.**

Sometimes ball-dropping is strategic. To train your partner to catch it. Other times ball-dropping is genuine incompetence. Or just life. Mistakes happen. Often, it's really difficult, if not impossible, to tell what's strategic and what's not.

If she wanted more parity, Rena observed, she should probably "just let him do it, be okay if it gets bungled." But it's a bind, she pointed out. If you say, "I'll just do it," then you add something else to your endless list. But if you hand off the task, it can be more work. "He just takes his time with it," Rena said, and sometimes "he will actually forget." So the task still occupies space in her mind as she's wondering whether he'll do it or whether he'll forget or mess it up somehow. In this way, the item stays on her to-do list; it's an "open loop,"[13] but one Rena can't actually close, since the task is now on his plate.

In my experience, finding balls to let drop and finding ways for them to drop in a different place feels like a special kind of martial art. It takes a lot of training (for many of us, that means therapy) and a lot of planning (admin).

Ask this question: Can you let it fail? This is not a recommendation to let balls drop in a haphazard or careless way; rather it's designed to help you find your own bottom line. *What is too much for you?*

Practically, this can be a guidepost for what areas not to hand off—where you truly can't tolerate someone else's mistakes—and what areas have more wiggle room. Emotionally, the question presses deeper: Is there a limit in you that is so important that you can't tolerate its violation? If you can find it, then you may not need so much planning. Your resolve may do the work for you.

7. **Give the gift of admin.**

If generosity is more your style than negotiation or martial arts, try giving admin gifts. You could do it just to make the other person happy. Or, more strategically, you could do it to inspire reciprocity.[14] Research confirms what common sense intuits: good deeds can spur cycles of happiness, gratitude, and giving back.

The Rolls-Royce of admin gifts would be an admin vacation. Imagine offering a loved one an admin-free mind for several hours or days as you take over the phone or email, the snail mail, the planning of meals, the coordinating of any help from family or others—all the command and control aspects of this person's life. A very capable Super Doer might not welcome this gift, might not want to relinquish any of the usual excellence of his or her admin-doing. But a Reluctant Doer who receives such an offer in earnest might well melt into a puddle of appreciation.

The practicalities of giving an admin vacation are likely to be challenging, for all the reasons admin is sticky and harder to transfer than regular household chores. You might choose not to shoot for the stars, at least not at first. Underpromise and overperform. Try something small. Order an extra phone charger or headset for a partner who keeps losing his. A little admin can go a long way.

8. **Express appreciation—enthusiastically and promptly.**

It's not easy to be an Avoider and keep your partner happy. But it's possible. Some Avoiders whose spouses don't resent them are just plain lucky. The rest make appreciation into an Olympic sport.

Olympian appreciators don't just say thanks. They say, "Thank you so much; that's awesome you took back the old cable box." A psychologist I consulted recommends a further enhancement: Name the suffering involved. As in, "Thank you so much for handling the health-insurance stuff; I know it's such a pain in the ass."

An indirect way to show your appreciation is to respond promptly to admin-related requests. The day I interviewed him, Marcus said the "main thing" on his admin to-do list was "an email sitting in my inbox from my wife on luggage options that I have to make some decisions on." I didn't ask how old the message was, but any email that's "sitting" didn't just arrive.

Does this mean you have to let your train of thought be interrupted all day long by life admin just because your partner is doing it at the moment? No, not in my opinion. You can pick a time when you'll respond and ask your partner if that's soon enough—or ask when your answer is needed.

And if you're the one who wants to be appreciated, try admitting it. You can be playful—though avoid sarcasm. You might say: "Okay, here's a role-play idea. You be the HR person, and I'll be the worker who did a lot of admin today, which I'm going to tell you about so you can give me some sunshine. Game?" Or if that's not your style, just say it directly: "I'd really appreciate thanks for some things I did today. They are X, Y, and Z." Simple, yet rare.

## Troubleshooting

We've already noted a number of potential troubles, including ball-dropping and admin judgments. Here are a few more admin traps to watch out for, starting where we just left off—with the seemingly simple matter of saying thanks.





1. **Watch out for the gratitude impasse.**

Diego and Marcella love each other, but they are at an impasse. She's doing valuable things (like taking the car in for registration renewal, which is already overdue). He's doing valuable things (like building a shed in the backyard with their three kids). They disagree over which task is more urgent, more necessary, more valuable. They agree the car thing is more painful. But is it more pressing than building the shed? If they all died tomorrow, wouldn't the work on the shed with the kids be more valuable? But since they aren't going to die tomorrow, doesn't the car thing need to be done first? With all these undercurrents of disputes about the existential value of admin, which surfaced in our interview, Diego and Marcella are finding it difficult to say two words: *thank you.*

Have you been there? Where it feels like there are so many invisible things for which you should say thanks to someone—your partner, parent, friend—that you no longer know how to begin? I have. It's a heavy, hollow feeling. Like *Thank you* will just be words because you're not feeling it right now. Which can lead to waiting longer to say it, and then the backlog grows, and so on.

"Economies of gratitude" are complicated.[15] Sometimes saying *Thank you* can seem like a concession to the other person's view that something needed doing or to be done their way. But you can thank someone while also acknowledging differences of opinion. This may sound stingy, but it need not be. Consider "I know we disagreed about how to do this, but I really appreciate the time you put into it." What else can you do to move past the impasse? Try writing instead of speaking. Put it in an email. Pick one thing you appreciate and start there. Or give a small gift to say thanks.

2. **Watch out for planning-meeting debt.**

Danielle introduced me to the phrase *paper debt*, which I took to mean piles of paper that are so backlogged that starting work on them seems beyond hope. A similar thing can happen with joint planning meetings. The cause may be simple: Life gets busy and you don't have a meeting for a while. Then you begin to feel like any meeting will take seventeen hours, so you push it off further. First one person opts out, then the other. And now you're way behind—and possibly blaming each other.

One idea: Hold a very short meeting with a timer set to end it. Figure out the highest priority items. And give yourself a great reward, whatever that might be for you. And schedule the next meeting.

A deeper cause might be your structure. If one person prefers solo work and the other wants to work side by side, you may be pulling in opposite directions in these meetings. Try this strategy: Review the Admin Pleasures Inventory in Chapter 9, circle each person's preferences, then experiment. Try one person's method first. Save a few minutes to talk about what worked and what didn't. Next time, try the other person's method. Compare. Rinse. Repeat.

3. **Watch out for "Just tell me what you want me to do."**

This sentence can be said in two ways. In one, the eager gopher, in his first day on the job, says to his new boss: "Just tell me what you want me to do!" You know he'll do it immediately and enthusiastically.

In the other version, an irritated person who doesn't want to help rolls his eyes and says, "Just tell me what you want me to do." If he does it, he'll do it halfheartedly—and he may make you feel bad for asking.

The second version is the one to watch out for. This sentence is a demand that you solve the problem and give the other person a neatly packaged corner of it to do—or possibly that you should just give up asking for help at all. If you see this one coming, you can resist the pull to rack your brains for a discrete task and instead try telling the person what would actually be helpful, even if it's amorphous—like coming up with some ideas for how to solve X problem. If no helpful ideas are forthcoming, at least you didn't waste your energy trying to find the person a doable job.

4. **Watch out for "We need to do X," otherwise known as the imperial-delegation gesture.**

   Sometimes "We should do X" really means "*You* should do X." This is like the child who stretches her arm out, tissue between two fingers, waiting for you to snatch it up and throw it away.

   Again, a matter-of-fact response seems best. "Who should do that, and by when, do you think?" It's far harder to say, "You should," than to imply it.

5. **Watch out for adminimizing and admaximizing.**

   People in relationships polarize. Remember *Annie Hall*: For her, they have sex "constantly," three times a week. For him, it's "hardly ever," three times a week. You may become idiots and maniacs about admin-doing, polarizing around how much admin you each do. You might also polarize around your overarching attitudes to admin. In other words, you may become adminimizers and admaximizers in your positions on how much of your relationship time admin should consume.

   The adminimizer says: *I want this part of life to be as small as possible*. The admaximizer says: *Admin isn't a burden on marriage. Admin* is *marriage.*

   Be wary of absolute devotion to either position. Can you really make this the tiniest part of life and still be investing in your relationship and your household? Alternatively, can you really have a meaningful relationship with the other person if this is all you talk about? In the abstract, we can see that neither position is likely to be viable on its own; rather, we can see that adminimizing and admaximizing might be useful strategies in particular moments. In reality, in relationships, we often fall prey to one pole or the other.

Try to make each of your admin personalities assets for the team rather than the raw material for weapons and grudges. You may *feel* far apart whether or not you actually are. Even among the closest of personalities, differences emerge.[16] Figuring out how to appreciate those differences is important for any pairing of types.

In the next chapter, we'll consult the four admin personalities to draw out each one's unique wisdom. We'll cull their insights and learn from their strategies.

Mary, who participated in a brainstorming group, grew weary of discussing admin in relationships. She eventually asked: "If you don't have anyone to share admin with, how can you make it better?" The next chapter is for you, Mary. And for all of us who want life to improve, alone and together.

12

# Individual Strategies

Are we not of interest to each other?

—Elizabeth Alexander, "Ars Poetica #100: I Believe"

A better admin future depends on ideas for how relationships can change and how the world can change. Admin is not an individual problem, and its solutions are not primarily individual. Change is needed in law, policy, markets, employers, schools, communities, and families.

But what if nothing changes? Less grimly, let's just assume nothing has changed yet. I told you at the start that this is the book I wish I'd read ten years ago. Let's imagine it's ten years ago, and you and I are the only ones reading this book. The only changes available are those we can make ourselves.

An admin perspective can still help us. We can study our friends across the personality spectrum: the Admin Denier, the Admin Avoider, the Reluctant Doer, and the Super Doer. Though most of us are a hybrid of the types, we can imagine archetypes. We can learn from each of these personalities.

Each personality has strategies, and this chapter will lay out a few of them, summarized in figure 6.

Figure 6

**ADMIN STRATEGIES BY PERSONALITY**

| | Feeling Good | Feeling Bad |
|---|---|---|
| Doing | Super Doer<br>*Optimizing*<br>*Engaging others*<br>*Making time* | Reluctant Doer<br>*Illuminating*<br>*Minimizing*<br>*Transforming* |
| Not Doing | Admin Denier<br>*Rebelling*<br>*Escaping*<br>*Simplifying* | Admin Avoider<br>*Delaying and Deflecting*<br>*Appreciating*<br>*Trusting* |

Let's hear what the personalities have to say, beginning with the most challenging.

## Admin Denier

***"Not-doing and feeling pretty good about it."***

Recall the Denier. The Denier doesn't think admin is a thing. Or doesn't believe it needs to be, anyway. None of this is necessary, she says. If it even exists, he adds. The Denier is not-doing and feeling pretty good about it.

The Denier's strategies include *rebelling*, *escaping*, and *simplifying*.

*Rebelling*

Some people are Admin Rebels. Other people just rebel.

Chloe, a doctor, is an Admin Rebel in multiple areas of life. She once quit a job in protest over work admin. The paperwork burdens at the hospital that employed her were so overwhelming, they interfered with her ability to give proper care to her patients. She tried speaking out against the system. She tried evading it. The director of her department at the hospital prioritized doctors' staying on top of their paperwork over reducing paperwork. So she quit.

In life, she fights back against companies that waste her time —documenting their impositions and forcing confrontations. She doesn't always win, but she doesn't hesitate to put up a good fight.

Most of us don't have the time, energy, or chutzpah to be Admin Rebels. But sometimes we snap. I've heard many stories of regular people feeling overcome with frustration and pushing back. When we do this well, our rebellion has the potential to benefit other people in our wake.

What would inspire more rebellions? Anyone can snap, but protest may be more likely when a Denier or Avoider takes on a task usually covered by a Doer. (How does a Non-Doer become the designated Doer? Love, guilt, barter, and accidents are possible pathways.) Gender may have a role to play as well. Research indicates that men are more likely to change their behavior in response to wait times—for instance, forgoing giving blood if the wait time increases.[1] This finding suggests it may be worth sending a man in to do what has been a woman's job.

Admin Rebels. The world needs more of them. But rebelling can be time-consuming. What else does the Denier bring?

*Escaping*

Escaping is less confrontational, and possibly a better fit for the archetypal Denier. Refusing to check your email for stretches says:

*This cannot be urgent, so I am not going to let it into my world.* Or even, *For this period, the admin you send me does not exist.*

There are plenty of techniques: Turning off your phone. Or never getting a cell phone in the first place. Going off the grid for vacation. Refusing to give out your phone number or giving out a fake number to retailers that request it. Creating a spam email account and using that address for retail purchases or anyone whose messages you don't ever want to receive unless you go looking for them. Not checking your snail mail for a period; for some, like my interviewee Phil, that period is "three or four months."

Of course, admin-doing is required for some of these escaping techniques, like creating a spam account. But we are taking the Denier's motto and using it to craft strategies.

Being charming helps here. Escaping the usual demands isn't easy; doing so and being likable is a feat. Perhaps for this reason, Deniers are often charming. One of my interviewees was explicit about using charm as a substitute for admin. But most seemed unaware of this technique—not so surprising if they don't see admin as a thing. If admin were visible, though, we'd all see its impositions, and the privilege of rebelling and escaping could be extended beyond the bold and the beautiful.

*Simplifying*

Choice is a major source of admin. Friends who've moved to the United States from overseas describe grocery shopping in this country as overwhelming: How does anyone decide among so many brands of laundry detergent? And empirical studies suggest that all those choices aren't necessarily making people happier; indeed, they're making at least some of us less happy.[2]

One response is to opt out. Decide none of that admin needs to be done. Resist the pull of social pressure to keep up with the Joneses. You might choose the nearest doctor, gym, kindergarten, and let whatever happens happen. You might forgo the whirl of camps and afterschool activities to just be with your kids, as Lillian did after her divorce. Or, if you have the money, you might even buy your way out of the fundraiser you were supposed to organize, as Lillian also did, during the admin onslaught of her divorce, by donating the amount raised last year. Or you might take more global steps.

My interviewees Elsa and David moved to Sweden—land of free, high-quality childcare, schools, and health care—in part to simplify their lives. Elsa also has smaller simplifying strategies. She buys the family's clothes in black and gray. "This may sound insane," she said, but "that way I don't need to expend mental energy on putting together 'outfits.'" Their menu is the same each week; for example, Mexican on Mondays, salmon and broccoli on Tuesdays, and so on. She has stopped buying birthday presents for people, and the family does not subscribe to any magazines or newspapers. "With every potential purchase—be it a car or a toaster oven—I ask myself whether we really need this item, and I consider the time and energy I will need to spend maintaining it or discarding it if it breaks," Elsa said. So when they need a car, they rent.

Some people are natural simplifiers. Their things and their homes are spare and orderly. If you're not there already, I cannot advocate, from an admin perspective, undertaking the transition costs to adopting their approach. Simplifying is a strategy that can bring the Denier full circle to meld with the Super Doer. If you are not inherently minimalist about stuff and you have successfully "Kondo-ized" your life, you are, most likely, a Super Doer.[3]

Some non-Deniers may be thinking, about the Denier's simplifying, *I wouldn't want my world to be that spare.* Or about the escaping, *I wouldn't want it to be that quiet.* Or about the rebelling, *I wouldn't want that much conflict.*

The Super Doer has the most to say in response to the Denier.

## Super Doer

***"Doing and feeling pretty good about it."***

The Super Doer finds the Denier baffling. How is it possible to not check your snail mail for three or four months? Three or four days, perhaps, if you're traveling. But months? This is incomprehensible to the Super Doer.

And yet there is no one single Super Doer. A Super Doer may do a lot of admin or a little. He may be hyperorganized and enjoy keeping most admin in his own capable hands. Or she may outsource much of it, leaving her with the admin of supervising the direct Doers, paid or unpaid.

The Super Doer has many admin strategies. We'll focus here on *optimizing*, *engaging others*, and *making time*.

### *Optimizing*

Super Doers tend to have effective methods for getting things done. This section will relate a subset of the specific methods Super Doers have described to me. But first, an important caveat.

This strategy is called *optimizing*, not systematizing. Super Doer methods are not uniform magic bullets. Rather, they suit the particular individual. Some Super Doers are highly systematic about their admin; others have just found a good way to live with doing it. I wish that studying Super Doers had led me to the golden monkey of admin success—whatever that might mean to any given person or even just to me—but it has not.

Indeed, one thing I gathered from interviewing Super Doers is that strenuous striving toward better admin systems is not typically a feature of their lives. Erin put it this way: "With some exceptions, like a calendar, which does help you, basically the key is you have to get organized. I don't think there's many tools that are going to help you. . . . The people who are endlessly buying these things—they're not organized. And so they're looking for something."

Here are some other specific methods and tools deployed by Super Doers (and their close cousins) to optimize their admin-doing:





- Online bill pay and autopay (with reminders to check the bank balance before big payments)
- A repository with shared access (through Dropbox or Google Docs or similar services) for important documents, including IDs and passports and kids' immunization records
- Shared calendars, for instance, through Google Calendar (loaded into everyone's smartphone)
- Amazon app for household supplies (for bypassing the to-do list and ordering from a company "where the customer is always right," at least for some[4])
- Scanning and shredding (which one interviewee aptly called "up-to-date filing")
- Taking photos of information (instead of scanning or writing information down)
- Using apps like Venmo or Square Cash to push money at people
- Sorting mail while walking in the door and then recycling or shredding promptly
- Bypassing the to-do list by dealing with requests on the spot (like texting a recommendation for a dentist while the asker is still there to ask questions, provide her phone number, etc.)
- Keeping a stockpile of gifts
- Keeping a travel bag ready for trips (if you take a lot of them) and setting out a specific bag for an upcoming trip (to drop in things when you or others think of them)
- Renting or leasing your home instead of buying (particularly if you're fortunate enough to afford a full-service rental)
- Managing helpers, paid or unpaid, depending on resources and community.

The last is a strategy of its own.

*Engaging Others*

Super Doers engage others skillfully, recognizing that life's work takes a village—or an army.

Super Doers at their best seem to make an art form of interactions around admin. They are firm, humane, engaging, determined. Not imperious. They may ask to speak to a supervisor, but the aim is not to get revenge on the first unhelpful desk clerk; it's to get the job done effectively and personably. Perhaps Super Doers are more often keyed in to the effectiveness of treating people in admin interactions as human beings because they're more likely to see admin itself as part of the human sphere—rather than some horrible alien imposition.

Not all Super Doers outsource admin, but of the people I've met who outsource, a high proportion are Super Doers. Outsourcing could, in principle, be a strategy employed by any admin personality with adequate resources or influence over others. So why are so many outsourcers Super Doers?

Finding someone to do your admin is typically no easy (admin) task. For this reason, many people with the money outsource housecleaning, but few outsource admin. Even my Super Doer interviewee Hazel, who had in the past *hired* a personal assistant (PA) and had also *been* a personal assistant, could not face the trouble of finding a new PA. She describes the month when she had a personal assistant, whom she found through serendipity, as "one moment in my life that was fabulous." Nonetheless, Hazel can't handle the prospect of the search, "the risk of bringing somebody in that isn't the right fit" and the "ramp-up time of orientation."

Money is part of the story, of course. Before deciding you can't afford it, though, you might make a list of any unclaimed rebates, reimbursements, unreturned broken items, and FSA or insurance submissions on your to-do list and consider if a few hours of personal assistance every month might bring in enough money to cover a PA's wages.

While researching this book, I have been asked several times for advice on how to find a personal assistant. What I can tell you is what I have seen. The largest, most accessible market for PAs is for virtual assistants. There are now lots of companies providing remote personal assistance — businesses such as Zirtual and Fancy Hands and Habiliss — and so the admin-startup costs are pretty low. But no one I've met has had a great experience with virtual assistance.

When I tried it, while writing this book, most tasks had a physical component that broke the chain of what the virtual PA could do. I anticipated this problem and chose a company whose online reviews said its assistants could outsource physical tasks to TaskRabbit, but my particular assistant had no experience with that. Moreover, geographical distance (just across the United States) meant a cultural divide that we couldn't easily bridge, which is relevant even to simple things like designing a babysitting flyer or choosing a restaurant. Turnaround-time delays also meant that handing her tasks didn't close the loop in my mind. I could have complained about the delays and the culture gap (and presumably the TaskRabbit virginity!), and the company would have found me someone else. But ultimately I wasn't convinced I had enough work that was truly virtual to make the additional interactions — and the substantial cost — worth it.

Nearly all the effective PAs I've heard about were found through personal contacts. A few people found their PAs at work. My interviewee Elisa works in a law firm where the lawyers are encouraged to give their life admin to their work assistants, who, she reported, understand that's part of the job when they're hired. Elisa finds that help to be a lifesaver, but, interestingly, I also saw in my admin travels that not everyone wants that kind of help, even if they can afford it. My interviewee Anabel cast boutique firms that provide comprehensive admin assistance as a "dystopia": In her view, these businesses try to "take care of everything," so that their most valued employees "can do work that's of higher and better use as defined by the organization." The firm can arrange backup care for your sick child, make travel plans for your vacation, and buy your spouse an anniversary gift. Their goal is to "eliminate all admin and treat things like relationships *as admin*," so that the firm can live your life for you—and so you can work long hours and travel when asked. These admin-doing initiatives by firms are usually praised and appreciated for addressing the needs of working parents, and I think that praise is well deserved. But talking to Anabel helped me see how that kind of assistance, particularly in the context of a 24/7 work culture, could be unappealing to some.

On-the-job assistance with life admin is relatively rare. More people I've talked to who have successfully outsourced life admin found help through their existing *personal* networks. Most commonly, PAs were something else to people before they became their PAs; they were their dog-walkers or babysitters or unemployed friends or grad-student neighbors. That personal connection translates into a belief that the person is trustworthy and intelligent, possesses the relevant skills, and knows your town. And if the person already helps you with something else—like taking care of your pet or your kids—the information-transfer at the outset is minimized. (If you're considering this, but it seems awkward to ask someone in your life if they want to be a PA, one idea is to draft a short ad and ask the person to forward it along to anyone who might be interested.)

In short, unless you already have a prospect nearby, it generally takes a fair amount of admin to find help with your admin. No wonder that it's mostly Super Doers who get that finding-help-with-admin admin done and create an outsourcing situation that works. Outsourcing admin doesn't always cost money, though. Some talented people manage to involve relatives or friends for free. For instance, the mother of my interviewee Tina takes care of figuring out and ordering whatever clothes Tina's kids need for the next season's weather.

Engaging others enables Elisa's approach: "I make lists about what has to get done. And then I try to always delegate whatever I can delegate first, so that things can happen in parallel. That's my overall strategy for life. Have things happening simultaneously." Nice help, if you can manage it.





*Making Time*

One friend found a solution to her hatred of going through the mail. She would sit down on a Sunday afternoon and deal with it all. By setting aside time to do the task, she made it less onerous. Her story brought to my mind the image from *Bridget Jones's Diary* of a woman whom Bridget aspires to become: "I read in an article that Kathleen Tynan, late wife of the late Kenneth, had 'inner poise' and, when writing, was to be found immaculately dressed, sitting at a small table in the center of the room sipping at a glass of chilled white wine."[5] This sounds like an elegant way to take on an admin task. Of course, for many of us, buried under kid clutter or the detritus of a busy life, Kathleen Tynan's posture feels laughably out of reach.

What might sound more plausible is the image from my interviewee Shira of "succumbing." Shira has learned to yield to admin at moments in life when nothing else is possible, like when her husband, who has post-polio syndrome, fell and broke a leg while they were traveling on a cruise. What else could she do but embrace that every waking moment, when she wasn't caring for him directly, would be dedicated to figuring out and planning how to manage the situation and get them home? "I kind of gave myself over to it. *Hey, this is a crisis. Let's see what we can do with this.*" And then "I just sort of did it; I can't say I hated it."

Shira has also applied this approach of succumbing to the rhythm of her daily life. Over her years as a writer, she's realized that her most creative hours are not in the morning—when so many people recommend doing creative work—but in the afternoon, when her anxious morning energy has burned off. She therefore does her admin early. Other Super Doers I've met instead make time for admin later in the day, when their minds are tired from finishing projects they value more highly. Whatever the timing, for admin that needs doing, the Super Doers have something to teach us about, as Lauren put it, "letting myself enjoy admin."



## Admin Avoider

***"Not-doing despite feeling (more or less) bad about it."***

The Avoider does not like listening to the Super Doer. Hearing about all that the Super Doer is doing and how effectively is not fun. Whether the Avoider feels only a little guilty or feels actually ashamed, she does not enjoy the Super Doer. Except, perhaps, if they are married (and not always even then).

Feelings aside, the Avoider also has a number of great strategies in her toolbox. Here we'll consider *delaying and deflecting*, *appreciating*, and *trusting*.

### *Delaying and Deflecting*

Delaying and Deflecting are close cousins of the stereotypical Avoider strategy—Ball-Dropping. But the first two are less risky if you're flying solo or can't assume a particular person will pick up what you drop.

Delaying can involve techniques as simple as not responding quickly to group email messages so someone else responds instead. Or until the asker figures out the answer on her own. My more tech-savvy friends use this on me sometimes; I write with some computer problem I can't figure out—probably something obvious to most readers—and my addressee (who I know checks email often) doesn't reply quickly. And often enough, twenty minutes of Google searches later, I've found an answer. This is a useful tactic when you want to say this in response to a request: *What I would do to figure it out is the same thing you would do to figure it out yourself.*

Deflecting has a slightly more active energy. When someone attempts to hand you admin, see if you can hand it back. I've been trying this out.





> The dentist's receptionist holds out a receipt for my visit. I look at it. She stares at me, wondering why I'm not reaching for it. Slowly I produce the words. "Is there any way that your office might submit that directly to the insurer?"
>
> "Sure," she says. "As long as we have all your information on file. Which"—she glances down at my chart—"it looks like we do. You're all set."

Who knew this was possible?

It doesn't always work, of course. Sometimes the receptionist says, "No, that's something you have to do." Sometimes unpleasantly. But I didn't even think to ask this until I began paying attention to admin. I assumed that all such offices had their practices and that if they could be doing things in a way that was easier for you, they would be. Especially for those of us fortunate enough to have private insurance and a choice of doctors.

But it turns out that if you ask, doctors' offices and others may take on projects that would otherwise suck up your time. First you have to see the admin coming. To me, admin about to land feels like a piece of paper that is too heavy in my hand, like an item in search of its list, or like a mild sensation of dread.

Admin comes flying in the intimate sphere as well. The friend who asks, "Could you give me the recipe for that amazing broccoli appetizer?" The partner who texts, when you both have equal access to the automobile records, "Do you know our car registration number?" The ten-year-old child who asks, "Where's my hairbrush?"

Do you immediately start a mental search for the hairbrush, start digging for the records, or add "send broccoli recipe" to your to-do list? Whether you do these things is your choice; my hope is that you see in those moments an opportunity to do or to deflect.

Deflecting isn't always easy, even if you see the admin coming. A simple *Sorry; too much trouble* probably won't be a welcome response to the request for a recipe. *Find it yourself* is unlikely to make for happy partners or kids. Even less so the sarcastic retort *Do I* look *like your secretary?* Probably the worst option is the sarcastic retort alongside your doing the work anyway. (The tech-savvy version would be to search the person's question on the snarky site LMGTFY.com—Let Me Google That for You—which generates a link that you can send the person showing how to Google the question.) You've won no goodwill—indeed, you may have spurred hostility—and you've rewarded the person's request.

With family members you might send it back gently or playfully (as in, "I bet you'll find it before I could!"). Or you might decide to do the task on this occasion, but deflect on behalf of your future self (as in, "I'll find the car info this time, if you follow along and agree to learn it for the future"). And remember the sage advice that in close relationships, you should be prepared to say something three times before you resent the other person's not changing their behavior.[6] Habits die hard. Especially habits that are handy for others.

It's like the famous parable, credited to Jack Warner of Warner Brothers, about managing people at work. It goes something like this: Every person who comes into your office arrives with a monkey. If you let him, he'll leave it on your desk. But you're the zookeeper. Your job is to keep the place clean. When the person leaves, make sure he takes his monkey with him. "Otherwise at the end of the day, you'll have a screaming jumping troop of monkeys and monkey shit all over the place."[7]

Finally, note that seeing admin before it lands can also help you help others. I confess that I was the one who wanted the broccoli recipe mentioned above. I asked my stepmom how she had made broccoli taste so good. She started to tell me, but I stopped her, realizing I would never remember. But did I need her to write the recipe down? My admin radar kicked in and I asked to make a video of her explaining it. I think we were both delighted. Admin deflection isn't always selfish.





*Appreciating*

Deniers charm; Avoiders appreciate. This reflects an archetypal divide between Deniers and Avoiders: saying thank you. Whereas the Denier doesn't think admin is a thing, the Avoider sees it, detests it, and runs the other way. If someone else intervenes and heads it off at the pass, the Avoider is likely to be quite grateful.

Saying thank you is not always easy. Saying thank you well is even less so. The previous chapter, on relationships, explored appreciation in depth, so we won't dwell on it here. But remember that an effective thank-you is specific and connected. And keep an eye out for the gratitude impasse.

*Trusting*

Another Avoider strategy is trusting. Trusting that things will work out. Trusting that the neighbors around your new home know when to put out the recycling so you can just follow their lead. Trusting your own intuition—about, say, the right price point on an apartment you want to rent or the safety of putting plastics in the microwave or the right foods to feed your child at her age—so you don't need to do further research or pondering before making a decision. Trust, the Avoider teaches us, equals less Cover Your Ass admin. In the words of one interviewee, "I will overthink when my instincts are insufficient."

For some of us, of course, trust is a luxury beyond reach. A mere handshake may protect a person with racial or economic privilege, whereas a person who wields less power may need formal documentation.[8] This is well known to Jerome, whose legal advocate told me the story of how he lost his home. Jerome lived in public housing with his father. Since Jerome earned $300 a month in public benefits, he qualified for the housing, but only his father's name was on the lease. For years Jerome asked the property manager for the simple form to add his name. The property manager put him off, repeatedly saying, "Ask me later," and eventually, "It's fine; stop bothering me." But when Jerome's father died, it wasn't fine. The public-housing authority evicted him as having no legal tenancy. So trust isn't always an option—or a good idea.

And some of us don't feel capable of trust, even if the world is on our side. Note, however, that trust as a *strategy* doesn't necessarily require having a trusting *feeling*. On the contrary, some Avoiders avoid precisely because they are anxious and the amount of research they would require to feel confident seems so inconceivably vast that they can't even take the first step. I feel this way about some food-safety matters—like I don't even want to begin to know, because what I might uncover is too much.

Everyone's risk tolerance and safety net are different. If you think you've lost your credit card, do you call promptly to cancel, and start the admin cascade, or do you wait to see if it turns up, checking (or not) for any unauthorized purchases? The Avoider teaches us to see the tradeoff between trust and Cover Your Ass admin.

### Reluctant Doer

***"Doing despite feeling not so good about it."***

The Reluctant Doer likely longs to be one of the other characters. He may wish for the carefree state of the Denier or even just the additional free time enjoyed—notwithstanding some possible guilt—by the Avoider. She may aspire to be the Super Doer, somehow managing to stay enough on top of things to feel pretty good about admin. But the Reluctant Doer is, for the moment, stuck. Doing and not wanting to be doing. Yet not seeing another way. And so she has her own strategies, which include *illuminating*, *minimizing*, and *transforming*.

*Illuminating*

The Reluctant Doer tries to make admin visible. This might ungenerously be called *complaining*. *Illuminating* has a better ring to it.

More seriously, the project of trying to make this work visible so it counts is no small undertaking. People have told me many stories about their efforts to make admin visible—making a jobs chart on the refrigerator and including admin tasks; sending email updates to a partner to prompt a thank-you; inviting a trade with some other task, to name a few.

Some Reluctant Doers have made strides to make admin visible to themselves—to recognize it and resent it less, or to break it up into smaller, neater pieces. One talked about being his own personal case manager, breaking off discrete tasks and assigning them to himself on particular days.

Most of us muddle along. But seeing admin as a thing has helped me feel, at the very least, less confused and guilty when I lost whole days or weeks to it or lost my focus to the parallel shift. Casting a spotlight on admin has also helped me to devise and try other strategies.

*Minimizing*

Minimizing is different from simplifying. Simplifying, a Denier strategy, means making significant life choices that eliminate categories of admin. So for the Denier, it's possible to say that the admin really doesn't exist. Move off the grid, don't have kids, sever ties to extended family. These kinds of ideas work for a bold few. Most of us aren't prepared to organize our lives around eliminating admin, though, at least not in major categories. We want to live where we want; we want the relationships we want; we are prepared to face some admin consequences.

Smaller steps can make a significant difference in how much time admin takes, however; this is adminimizing. Anticipating the admin involved in an activity may help you see, first, if you really want to do the activity at all—remembering Stephen Covey's admonition that "you are always saying 'no' to something." And if you do want to do this activity, then looking through an admin lens may help you choose a low-admin version.

Socializing at whatever age can be high- or low-admin. Kid playdates can be an admin nightmare. Under the model of parenting common these days in well-off families in the United States —a model that the sociologist Annette Lareau calls *concerted cultivation*, wherein parents work to shape their children's potential rather than trusting their natural growth—kids' schedules can be more complicated than their parents'.[9]

One summer, I struggled to arrange a playdate with a little boy my daughter had liked during the school year. This boy had a very full schedule, which didn't match ours, and eventually I gave up on trying to arrange that particular playdate. Instead, I concocted a low-admin, high-reward play event. I just picked a convenient date and time on the playground, checked for availability with one family, and sent an email to both my kids' whole classrooms (on a preexisting list), saying we'd be there then. I didn't ask for RSVPs; I didn't bring much in the way of snacks (some juice boxes). We just showed up, and so did a bunch of other families.

My kids were thrilled. Various parents thanked me for organizing it. And yet, to me, this was easier organizing than most regular playdates. Not least because if someone can't make it, you don't have to adjust anything.

This can work as well for some adults. It's the equivalent of a Brit's "I'll be in the pub from seven." Close friends already know which pub, and whoever can show up does. People do this across many contexts—religious gatherings, after-work drinks, book clubs, and more. Low-admin socializing.

One other kid example is telling. (Feel free to skip if you've opted out of kids and kidmin.) Offering your children creative or meaningful experiences can involve a lot of admin or a little admin, depending on how you do it. There are a zillion craft books that claim to show you how to make a papier-mâché post office, build an actual treehouse, or produce a volcanic explosion. They usually have a long list of ingredients that are purportedly household items but that never seem to be in my household. The low-admin form of inspiring kid activity—be it adventurous, artistic, or otherwise—builds on something you already know and like, so the set-up admin is slight and painless. For me, that was learning poetry with my kids. For you, it might be giving them ukulele lessons (a miserable failure for me) or teaching them whatever game or skill or craft is already in your repertoire.



For admin that we can't reshape in our preferred image, like insurance submissions, adminimizing can mean containment. I have sometimes set myself rewards for doing half an hour of intended admin *and no more*, especially if I start admin-doing at night. (The solo Admin Study Hall protocol in figure 5 can help me.) Timers and rewards—gimmicky as they may seem—keep me from toiling away way past my bedtime, with diminishing returns as I get more and more tired. Late-night admin entails the added risk of making the next day a dim haze of an admin hangover, where all admin looms large and little of it gets done.

*Transforming*

If admin is a means to an end, something we value for the result but not the process, then we can also transform our admin. We can make admin more of an end in itself.

Sometimes the change is dramatic—admin is turned into something that no longer resembles admin. The writer Henri Nouwen described how he "once spent long hours looking in Dutch stores for a birthday gift for my father or mother, simply enjoying being able to give."[10] Compared to glancing at an online list of "best gifts for men/women in their 80s" and ordering the first one off Amazon, Nouwen turned what could have been admin into an entirely different activity. Nouwen didn't recount what gift he got or if he even found a gift at all. The event was the searching.

Most of us have neither time nor inclination to turn admin into soulful activities. But we are always making choices about the process of how we do it. The transformation is subtle, not categorical.

A wise person pointed out to me, as I began a search for a divorce attorney, that this kind of search, amid and related to a painful time of life, feels like a huge waste of time. Yet it is often vitally important, not just for its outcome, but for the experience. If you interview more than one person, then searching for an attorney—or a therapist or any professional helper—may prompt you to retell your story to multiple people. You may hear their thoughts. You will certainly hear your own thoughts. The process is a nightmare in part because it is an emotional as well as a pragmatic endeavor. Recognizing that the process has benefits doesn't make it easy. But letting go of the sense that we are wasting our time may help reduce the suffering.

Valuing the process may also help us make small changes in how we do it. I have often found that blocking in five minutes before a meeting or event, at work or at home, to consider my goals for that meeting—and, more important, whether there is a way to make that meeting meaningful in and of itself—can make a difference in how that meeting feels. Even just asking myself the question of my goals as I walk to the event, reminding myself why I'm doing what I'm doing, seems to ground me and make me more effective.

I try to do this in trivial ways with admin projects I dread. Doing quarterly tax paperwork for babysitters can be a tedious task. Remembering, as I sit down to do it, how fortunate I am to have the money and luck to employ someone who cares for my children so beautifully has helped me to move through that with a little grace.

Our discussion of admin futures started at the local level—first, with relationships, and here, with you as an individual. Reading some exemplary strategies for each personality should have rounded out your sense of these types and how they operate. So next time you encounter an admin problem, you need not remember any one strategy; you can simply ask yourself, *What would an Avoider do?* Or *What would a Super Doer do?* You can try on a different personality and see what changes.

The final chapter broadens our frame to thinking about structural change. It's time to ask, *How can admin get better for all of us?*

# 13

# Collective Possibilities

Oh my God, how many minutes are being wasted every day all over the world?
—Bella, brainstorming-session participant

Lately I have dreams about admin. Not literal dreams, where I'm worrying a scheduling puzzle or missing a deadline. I have admin fantasies—dreams about the world as I wish it were.

When I sat in that chair in my bedroom making a list of the invisible admin I was doing after my second child was born, I thought admin was only *my* problem. I thought it was a trivial but overwhelming thing that I had to fix somehow by myself. Now I know better.

You are not alone. And neither am I.

I therefore promised you a book that wouldn't add further blame and shame to your admin frustrations or tell you to organize or meditate your way out of your admin woes. I promised not to Zen all over you.

Through my admin travels, I have come to believe that reform can happen in law, government, markets, employers, schools, and norms. By telling you some of my dreams, I hope we can change our reality.

*Admin would be visible.*

Visibility is central to all my admin dreams.

In my admin utopia, a special device—an AdminOmeter—would track how much time you spent on admin in a day. Effortlessly. No more trying-to-figure-out-my-admin-problem admin.

Miraculously, the AdminOmeter could even estimate the percentage of your mental bandwidth that admin was taking up—much like a body-fat calculator—and at what hours of the day. Some might not want this information, so its use would be strictly optional. I picture the AdminOmeter as one feature on a sleek digital watch that can magically track your admin-doing, physical and mental, and feed the data into an app in your phone that you can keep private or choose to share.

The AdminOmeter would help make admin-doing visible in relationships, for those who wanted that backdrop to conversations about shared labor. And, yes, a related device would exist to cover time spent on non-admin forms of shared labor, like chores, so that comparisons and tradeoffs could be assessed and made. (This feature would have been helpful to me and my ex-wife, since she cared more about tidying and did much more tidying than I did.) The AdminOmeter would also allow us to show government entities and airlines and insurers and other companies how much of our time they were wasting.

The AdminOmeter would make admin visible in a literal sense. That's my dream, but it probably isn't going to happen anytime soon.

What seems possible is for admin to become visible in a social sense. If everyone recognized admin as a thing, and if we started to develop a keen eye for its costs—to ourselves and each other—then many changes would follow. For starters, we'd see the burdens we impose on one another. If we ask someone for a referral for an eye doctor, and the person says, "Oh I haven't seen the eye doctor in ages, but I could maybe dig up her information over the weekend," we'd know that the person means *that admin request is a hassle.* We could decide how important the referral is and either say, "Never mind" or "Thank you thank you thank you." We'd also make informed choices about couples and admin—not assuming that one partner is the admin Doer; not assuming that the apparent Doer wants to be doing. When in doubt, perhaps we'd even ask if both partners want to be on admin emails or, if just one, then which one.

In a society where admin is visible, our interactions with companies would also be very different. Let's start with the fantasy version.

*Companies would give you your wasted time back.*

The year is 2029. Due to a (highly unusual) glitch in its system, Verizon has sent me a notice stating that I have not paid my bill for the past several months and that I now owe $147—even though I thought I had set up autopay three months ago. I spend an hour trying to create an online account, unsuccessfully, because the website won't recognize the combination of my account number and zip code. Maddening. Finally, I just pay the bill through the payment portal—which doesn't require an online account—to avoid a dent in my credit rating.

A week later, I learn that autopay had in fact been paying all my bills, so the paper bill I received was a mistake. At this point, I've overpaid by $147. But overpayment is not my primary concern; I will get that money back once I find time to complain. My concern is with the two hours I will have wasted.

In my imagined year 2029, I write up my complaint and send it to Verizon. I receive this response immediately:

> We are terribly sorry for wasting your time in this way. Please accept, in addition to our apologies and a refund, the return of your wasted time in the form of a personal assistant, Charlie. Simply click the link to choose the two hours when you would like to

schedule Charlie's help and indicate whether you prefer his assistance to be remote or in person. Charlie will be at your service during that time, ready to accomplish any of your admin tasks.

Yours truly, Verizon

When Charlie shows up, he is utterly competent and perfectly private. The same Charlie comes whenever I receive a time refund, no matter who wasted my time, and so he retains information about me—to whatever extent I like—which makes delegating admin to him seamless and easy.

This sounds like science fiction. And it is. But there is a realist corollary.

We should be able to recover money damages for lost personal time—getting a rough dollar equivalent of our time returned to us.

Under the law in the United States—and in most other places, as far as I can tell—individuals generally cannot claim damages for lost personal time due to breach of a contract, even when the lost time results directly from another's breach.[1] In one striking case, a court declined to grant recovery for all the admin hassle that accompanied a grocery store's security breach affecting the plaintiffs' credit cards. The court explained that the time and effort these individuals had to spend talking to banks about fraudulent transactions and changing their billing arrangements were just "the ordinary frustrations and inconveniences that everyone confronts in daily life with or without fraud or negligence."[2] In other words, the admin of life simply doesn't count.

Yet businesses can claim damages for employees' lost time, so by extension, individuals should be able to recover for their lost personal time.[3] Such recovery is not beyond imagining. In the UK, a Financial Ombudsman Service has been empowered by Parliament to award damages to customers who complain of "material inconvenience" imposed by banks and credit card companies.[4] And in the United States, the Identity Theft Enforcement and Restitution Act requires convicted identity thieves to "pay an amount equal to the value of the time reasonably spent by the victim in an attempt to remediate the intended or actual harm"[5]—that is, the value of your time spent on identity-theft admin. And no new law would have to be passed for judges to begin to permit recovery for lost personal time where there is a breach of contract or violation of tort law. How should courts decide what our lost personal time is worth? The question's not easy to answer, and scholars have puzzled over it. Recognizing admin as a form of labor invites a novel answer: Hours lost to admin-doing could be valued at the "going rate" for employing a personal assistant.[6]

Permitting lawsuits for lost personal time should not, and will not, lead most of us to file suit against the companies that top our admin most-hated list. (Who has the time?) But a few high-profile lawsuits on point—or awards of lost-personal-time damages tacked on to other lawsuits—could change the terms of exchange between customers and companies. A well-designed legal regime should prompt companies to provide *easy* means for compensation to avoid the threat of lawsuits.

Better yet, to avoid paying damages, companies would surely try harder not to waste our time in the first place.

*A Respect Our Time rating scheme would evaluate institutions for how they use our time.*

Online reviews can be very helpful. But these reviews rarely give us reliable information about time wasting. Ratings often conflate product quality with everything else. Even comments specific to customer service equate friendliness with time well spent. To my mind, if I even know how friendly a company's customer-service reps are, then the return has already taken too much time and effort.

We need a simple Respect Our Time (ROT) rating scheme that evaluates companies on this dimension. Companies such as Amazon and Zappos have built their businesses in part on making the process of returns as simple as possible—automatically accepting and facilitating returns electronically—which multiple interviewees described as vital to doing the life admin of procurement.[7] But many companies do not respect our time.

Imagine a company that sells vacuum cleaners—let's call it Schmoover. Schmoover has a one-year warranty on vacuums and parts but requires that the buyer personally take the broken vacuum to a service center to get it inspected and repaired—no option of shipping it, even if you bought it online. Schmoover's disregard for customer time is hard to discern before purchase, however. The lack of information about whether companies respect customers' time makes "comparison friction"—the difficulty of comparing products before buying them—particularly great for the time dimension of a product's value.[8] This is why we need ROT ratings.

Employers could be rated in this way as well. Lauren vented about her employer's wasting 80 hours of her time by changing her insurance coverage, without even an apology. Who pays attention to these matters when accepting a job? Perhaps if we had the information to compare prospective employers on this basis, some of us would.

Employers could rise in the ratings not only by avoiding wasting employees' time but by taking affirmative steps to protect their time. A friend who runs her own nonprofit recently took an admin vacation to address her son's health issues. She knew this was a benefit of being her own boss, a privilege most people don't have. But the Family Medical Leave Act, which requires large businesses to provide up to twelve weeks of unpaid leave (with no job consequences) for certain employees so they can care for themselves or their families, arguably covers unpaid time off to do the admin of caring for a family member with a serious health condition.[9] *Unpaid* leave may be little help, however, to someone attending to a serious health condition. Employers could go further and provide paid leave in these and other admin-related circumstances.[10]

A ROT rating scheme would ensure recognition for employers who took these steps and would allow prospective employees to make admin-informed choices about where they want to work.

Some companies need more pressure than a bad grade.[11] Here's where law can help.

*We'd have a right not to read the fine print.*

Do you read the fine print when you buy a household item? How often do you read even a single term before clicking *I have read these terms*? If you answered *never*, you are not alone.

Research suggests buyers very rarely read consumer contracts.[12] Not spending our lives reading fine-print contract terms that we cannot individually negotiate—and when comparable sellers may well have similar terms—seems a highly rational choice.[13] By one estimate, if all American consumers read the fine print on all the websites they visited for one year, the value of the time spent reading privacy policies would be $781 billion.[14] Or think of it this way: Would you want to hire someone for a job if you knew that person spent hours reading the fine print on the contract for his toaster oven? Would you think that person exercised good judgment in the use of his time?

Yet courts speak of a "duty to read" one's contracts. They lament contracting parties' failure to read.[15]

Contract law should abolish the duty to read consumer contracts.[16] Various legal doctrines have already chipped away at the traditional duty to read, but law should go further and protect people's freedom *not to read*.[17]

What would that mean? For starters, legislation could limit buyers' obligations under consumer contracts to terms that are displayed in something like a Schumer Box, which is a prominent area on credit-card agreements that must tell you key rates and fees associated with the credit card.[18] Anything outside the box would be interpreted in the buyer's favor. Sellers would be required to label the box to signal its unique importance—*Warn-*

*ing: Potentially Unfavorable Terms.*[19] An Unfavorable Terms Box for consumer contracts sounds obvious enough, but it's not the law.

Moreover, law should do more to tackle the problem of unenforceable terms. Sellers sometimes put additional terms in fine print even if those terms wouldn't be upheld in court. For example, a seller may include a warranty term that is shorter than the legally required warranty.[20] The company is banking on buyers' unwillingness to research their rights, much less go to court to vindicate them—and anyone who wants to fight back has to do both.[21] The freedom not to read fine print would also mean penalizing sellers for including terms that aren't enforceable.[22]

We need legal reform to prevent sellers from binding us, or making us think we are bound, to terms we don't have time—*and shouldn't make time*—to read.[23]

*The heavily regulated insurance industry would be held accountable for its handling of our time as well as our money.*

Dealing with insurance companies is many people's least favorite form of admin. Even among my interviewees who confessed a secret liking for some admin, none said they liked insurance admin.

People's wasted time is not necessarily an accident when it comes to insurance. Insurers have a financial interest in not paying out benefits. Outright denials of coverage need to be justified, but if the insured person just stops trying because the process is too taxing, that's a win for the insurer. It's no surprise that the term for lucrative admin drag—*rationing by hassle*—comes from the insurance industry.

No insurer will admit that it rations by hassle. And yet, since the insurer's financial interests weigh in favor of it, we would be naive to think rationing by hassle doesn't happen. Jay Feinman argues in *Delay, Deny, Defend* that insurers in the 1990s rethought their business model and created the modern claims machinery that uses time-consuming processes for individual claims in order to increase profits for the company.[24] According to Feinman, insurers investigate "excessively—by pursuing more and more information that is not really needed" to try to "exasperat[e] the claimant into submission," and "for those few who do not drop out . . . the final obstacle is a program of litigation that can wear down and defeat all but the hardiest."[25] (Or as Zack put it, "Basically they don't want to pay for it, and so they're going to make me pay for it with my time.")

Numerous reforms have been proposed, but these typically aim at getting people the money they deserve.[26] An admin perspective invites us also to focus on how much of our *time* insurers take. For private health insurance, people should know before choosing an insurer how much time, on average, participants in that plan have to spend doing admin to get a claim processed. Nothing in insurance brochures supplies this information. Insurance companies should be required to disclose this, but reform should go further.[27] Actual penalties should punish insurers who waste claimants' time—for private policyholders as well as public beneficiaries—even when the insurer eventually pays the heroic people who put up a fight.[28]

How insurers use insured people's time should be as important as how they use our money. Admin's invisibility means change in this direction is unlikely on its own; however, government's already substantial involvement in regulating the insurance industry makes this more than a utopian fantasy.

*We would have to face the ways our benefits regimes and our culture burden everyone with admin—especially the most vulnerable.*

A smart person asked me recently, "Where is the place on earth with the least admin?" Perhaps it is a Scandinavian country that aspires to take "perfect care" of its citizenry. Perhaps it is an ashram or retreat where the participants relinquish their devices and embrace silence. (In my version, that retreat center has a well-functioning front desk with employees who will come find me if an emergency arises at home in my absence.)

Going off the grid can help people feel freer, if they have the liberty to do so and if their obligations are well supported. But one need only listen to the (sometimes hilarious) monologues of our best meditation teachers to see that, to paraphrase a classic book title, *wherever you go, admin can follow you.*[29] The teacher Sharon Salzberg describes her experience of learning to meditate in India in her late teens and ending up spending the morning planning a simple trip to the market many times over. As she sat on a cushion in an environment with stunningly little stimulation, her mind just kept going over the plan. Her mind found its way to admin. Planning and scheduling are not problems unique to a culture with email or short attention spans. We cannot fully escape admin by moving elsewhere or by turning off our devices. Admin can follow us anywhere.

Yet context matters. Technology has helped in some ways, hurt in others. Escalating forms of communication technology offer us more ways to connect and more tools to support exercise and wellness, but they have also made us much more vulnerable to other people's admin demands. Self-serve admin has become the norm as we become our own travel agents, grocery-store cashiers, and more. Human resources departments have fewer people but more "portals." Technology has remade physical activities into virtual ones, so grocery shopping can become online shopping, turning chore into admin.

A country's social-welfare regime also matters significantly to the admin burdens on its citizens. I remember the first time I went to the doctor in England, where I lived in my twenties. I think I had some hacking cough that lasted long enough to warrant a checkup. I arrived at the office, waited a few minutes, saw the doctor, and left. It's been two decades since then, so perhaps I've forgotten some details, but I don't even remember filling out any forms. I know one thing for sure: No payment at the end, so no checkout hassles. No receipt to submit to insurance or to a Flexible

Spending Account. Nothing. I just walked out the door. It was disorienting, like floating.

Many Americans tell similar stories about receiving medical care in countries with a national health service. When you are used to the amount of paperwork and cost involved in the US medical system—the value of the time lost to physician visits alone is estimated at $52 billion per year—such moments are memorable.[30] Health care free of admin and other costs may sound idyllic until you remember that when you're interacting with a health-care system, it's usually because something is already ailing you. Under the current US health-care system, people often face admin onslaughts right when they are forced to confront the pain and suffering that come with vulnerable bodies and minds.[31]

Parents of young children are also hit hard with new admin just when they most need help. In my dreams, childcare and schooling would cover the hours when parents actually work, with affordable camps filling in the vacation and summer gaps. No more spending your few precious moments at home not dedicated to direct care, or your lunch break at work if you even get one, trying to find and juggle daycare, afterschool care, summer care, Presidents' Day care, and so on, because work hours and school hours don't coincide.

Childcare that covers work hours may sound fantastical unless you're familiar with the childcare system in a country like Sweden (where subsidized childcare for little ones is available from 6:30 a.m. to 6:30 p.m. and the school day is supplemented by before- and afterschool programs[32]) or France (where day camps housed in local public schools offer low-cost supervised activities and outings on days when school is not in session, and childcare services are commonly staffed with medical personnel who can treat common kid sicknesses[33]). Even the US government spends a billion dollars a year to fund an excellent and flexible childcare program (often covering fourteen hours each weekday year-round) for workers whose time it values highly: military personnel.[34] So while universal, free, and flexible coverage may be utopian, more modest versions already exist—even close to home.

If we truly saw the costs imposed by admin, we would also have to face that our poverty programs typically force recipients to jump through hoops that favor the most agile rather than the most in need. Some argue that rationing by hassle (for example, by increasing wait times) may be better than rationing through financial costs (for instance, raising copays) because poor people have more time than money.[35] But these theories are hard to prove, since the working poor may be paid an hourly wage with no paid medical leave and therefore lose money—or worse—if they have to spend hours doing extra admin, particularly if calls or meetings must happen at designated times that conflict with their work schedules (and if they don't have easy access to office equipment). And people with serious ailments like depression—individuals who may be especially in need of coverage or benefits—may also be especially disadvantaged by added hassles.

Seeing admin clearly, and recognizing how much it can hinder even people with privilege and support, should prompt us to rethink how we deliver benefits to people in poverty. Some innovative programs are doing just that: by creating one-stop shops for admin (for instance, integrating an application for higher education into a public-housing application or locating benefits offices at one site[36]); or proposing that benefits systems presume eligibility during a probationary or grace period (for instance, allowing a single mother to put her child in daycare while she does the admin of proving her financial status and starting her job search[37]); or scaling up grassroots admin help for low-income folks (for instance, supporting peer networks that help people solve their admin problems[38]). Other pioneers are working to develop technologies to deliver information and reminders at a time and in a manner that's usable to people who don't work in offices, and, since technology is no panacea, to press banks and government offices to maintain office hours that match participants' availability.[39] And lawyers and activists are laboring to make sure those technologies follow universal-design principles so sites and services are accessible to all comers, regardless of disability.[40]

Policy choices about health care, childcare, and other benefits involve considerations beyond admin, of course. And the switch to any new regime is likely to bring additional waiting and other admin costs in the short term, especially if the transition is bumpy or contested—as many of those who enrolled in a plan under the Affordable Care Act know too well.[41] But if admin were fully visible in these deliberations, we would see one more sizable long-term benefit to regimes that meet everyone's needs without hassle.

*Government at all levels—federal, state, and local—would look for ways to support admin-reducing technologies.*

Not all admin-reducing government innovations have a big price tag. Some happen through technological and regulatory changes.

Remember TripTiks? If you're above a certain age and grew up in the United States, you may recall this aspect of travel planning. Before a road trip, you could go to the local AAA Auto Club to wait while a live person pieced together a booklet of pages (spiral bound) mapping your journey from A to B. Your route would be highlighted (with an actual highlighter pen) to help you avoid any construction along the way. It seems inconceivable now.

What happened to TripTiks? President Clinton issued a decision in 2000 to order the military to "stop intentionally scrambling the satellite signals used by civilians," which increased Global Positioning System (GPS) signals tenfold.[42] This executive order triggered the development of individual GPS devices that reduce the time people have to spend planning a trip by car or on foot. The order made possible a world in which you can get in a car, type your destination into a device, and receive real-time directions on how to get there and avoid the traffic congestion.

Technology can also enable the seamless transfer of our information. My interviewees Elsa and David had a lot to say about these technologies, having recently moved to Sweden. The beauty of the Swedish way, Elsa told me, is that, once you're in the system, which takes some time, the system knows everything about you. Everywhere you go—banks, preschools, health-care providers—you merely enter your personal number, akin to a Social Security number, and the entity has all your information. This number lets you log in to government-information systems online or through an app on your phone to find out things like the status of your tax filing. The system knows when your children are due for their vaccinations, schedules the appointments, and sends you a postcard to let you know.

Back in the United States, Elsa and David had spent eighteen months contesting a $250 pediatrician's bill from a hospital visit after their first child was born. The pediatrician who examined and released their newborn son was affiliated with the hospital but not on their insurance, due to some combination of hospital error and Elsa and David's not realizing they'd have to call their own doctor to the hospital. "It boggles my mind," Elsa reflected, "that in the US every freakin' time you go to a doctor you have to fill out a form in your handwriting—with your insurance information and whether or not you have eczema. . . . It seems very Stone Age to me that we're still doing that. But luckily [David and I] moved away from that."[43]

In Elsa's words, it's like there is "a cloud hovering over Sweden with all of our information in it." While this could be a "little creepy," David added that "generally there's a much greater trust in government here than elsewhere, so that cloud hanging over Sweden doesn't seem to be problematic for people."

Entrusting all information to the government might be a lot to ask, at least in the United States, but in some states a friendly cloud of information is already accumulating—for instance, of registries of children's vaccination records.[44] And whatever you think about the government collecting new information on us, most people would embrace the prospect of government giving us back the information it already has. This brings us to tax reform.

*Tax returns would initially be completed by the entity that has the relevant information: the government.*

If you live in Chile, and your taxes are not complex, then each year the government sends you a draft of your tax return. You simply review it. Only if you reject the government's proposed return do you have to put in any real time or effort. Chile is not unique—far from it. As of 2013, eight countries offered a majority of their taxpayers pre-filled returns, eleven more made substantial use of partially pre-filled returns,[45] and many more countries have taken similar steps in recent years.[46]

To anyone whose experience of paying taxes involves pre-filled returns, the US system looks very strange indeed. US taxpayers work to gather the relevant information and run their own calculations, then submit them for review to the government, which typically had all the information in the first place.

From 2005 to 2006, California offered a pilot group of taxpayers state returns that were pre-filled with their information, just like in Chile.[47] The program was called ReadyReturn. Program evaluation of the ReadyReturn pilot calculated major savings for the state and the taxpayer.[48] The taxpayer's perspective is revealed by user feedback: One woman reported that she received a notice in the mail with her code numbers, then she reviewed her state-generated ReadyReturn on the website, checked it against her own records, and signed off. The whole process took fifteen minutes.[49] Other users enthused: "LOVE LOVE LOVE this service"; "This system rocks!"; and "THIS IS THE BEST SERVICE I HAVE EVER SEEN BY THE GOVERNMENT."[50]

Why didn't California's ReadyReturn program survive? You might think the problem was libertarians opposing pre-filled returns out of distrust of government. And there was some of that.[51] But the real story, most agree, is darker.[52] What happened to ReadyReturn was the result of lobbying and vested interests. Doing other people's admin can be big money. Intuit, the corporate parent to the popular software TurboTax—which asks a series of questions to help individuals who purchase the software prepare their own tax returns—spent more than $3 million on lobbying and political campaigns in the state over the five years following ReadyReturn's launch.[53] A member of the state tax board that had initiated the ReadyReturn program, Tom Campbell—a man who had extensive political experience, as a state senator, congressman, and budget director—observed that he had "never seen the public interest being overborne by private interests as clearly as it had been in this case."[54] What ultimately replaced ReadyReturn is a program that asks you questions to help you fill out your own return, much like TurboTax, combined with partial pre-filling.[55] Californians are still ahead of the curve, but ReadyReturn is no more.

At the federal level, the idea of pre-filled returns has been around since the Reagan administration.[56] Obama raised the idea in his first campaign, and key players in his administration elaborated and advocated the idea from then on.[57] More recently, Senator Elizabeth Warren introduced the Tax Filing Simplification Act of 2016, which would require the IRS to provide pre-filled returns to taxpayers with simple taxes.[58]

Could this happen? Sure. Will it happen? No guarantees. The federal government is already the rare entity that sees and quantifies and sometimes even monetizes individual time spent on admin. These calculations are required by the federal Paperwork Reduction Act.[59] Paperwork burdens imposed by the federal government on individuals and entities totaled 9.78 billion hours in 2015.[60] Tax preparation makes up a substantial portion of that—an estimated 1.855 billion hours in 2014—and for many households, this is probably the biggest paperwork burden they undertake for the federal government.[61] So admin visibility around paperwork is not new to the government, and every year at tax time, many people lament the absence of a simpler tax system.

And yet Americans continue to struggle to gather and record information—or pay or persuade someone else to do it—in order to pay their taxes, even if their tax returns are simple. Ridiculous. If admin visibility truly entered our collective national consciousness, perhaps this battle could be won. Some have argued that more radical change to our tax system is needed, and they are probably right.[62] But for starters, taxpayers with simple returns could stop spending time collecting and preparing and submitting to the government information that the government already has.

*Admin consultants would help us solve our admin problems and develop tailored systems for dealing with it all.*

I often wonder why there is no admin-consulting industry offering to help people fix their admin problems. I'm not talking about personal assistants (PA) who get paid by the hour to do admin for someone else; that market exists, though it's flawed in ways I've described, and it seems inevitably expensive if done well. And I'm not talking about high-end and specialized services that offer to help wealthy people organize their things or their lives.

I'm imagining consultants who zero in on admin as a beast requiring vast knowledge and expertise and who offer user-friendly packages to help people set up good systems for managing their own admin. The Admin Inc. industry I'm picturing would be an affordable option for a wide swath of regular people who can't or wouldn't pay an hourly wage for direct admin services but who might pay for a one-off consult—or splurge on it as a gift for someone else.

Admin Inc. would offer both tailored and off-the-rack solutions. I could consult with Admin Inc. on my to-do-list problem. An executive-function admin specialist—who knew everything about to-do-list approaches, high-tech and old-school—would ask me an intelligent series of questions to understand my needs, habits, and capacities (and my access to time and money). The rep would recommend the best system for me and then troubleshoot while I experimented with it.

Parents could order a graduation package for the child who's coming of age—whether at eighteen or twenty-one or thirty-one or forty-one—on a spectrum from tough love to gentle coddling. A Sink or Swim (SOS) package would offer children minimal help, perhaps a video and a packet of information on important admin realms of life that kids might overlook (renter's insurance, credit ratings), and explain the basics for the areas that parents will no longer manage (health insurance, taxes, cell-phone bills). At the other end of the spectrum, the Never Let Me Go package would offer aspiring adults a series of in-person meetings with a personal admin consultant who would explain adult admin, help set up systems to address it, and provide backup support by email, phone, and videoconference from then on. Admin Inc. would be the first port of call for the adult child who has accumulated credit-card debt (like Vera did, back in her Denier days) or years of unpaid E-ZPass car tolls (as did a young woman whose mother, Dorothy, I interviewed). Only then, if necessary, would that adult child go to her parents with a thoughtful proposal for their onetime help and a plan for avoiding a recurrence. Never Let Me Go would attract those monied folks with an affinity for indulging or controlling their adult children as well as some regular people who would mortgage their homes just to get someone else to have those tense interactions and take over worrying about their children's financial and bureaucratic well-being in the years ahead.

Admin-consulting services could be gifts for various life events. Friends and family could club together to give a newlyweds package or a wedding-planning package. How is the wedding-*planning* package different from the typical wedding *planner*? Rather than doing the wedding admin, the admin consultant helps the couple use this foundational event to turn an admin circus into a thoughtful basis for a life as partners. The Moving Experience package could bring expert understanding of admin's stickiness—and the full panoply of options for sharing or divvying admin—to those setting up a shared household.

What's your biggest admin-process failure? The possibilities are endless: Managing your email in a way that works for you. Deciding which admin is worth doing. Devising a system to organize your taxes and other paperwork. Finding the right app to do whatever you're trying to do. Ideally, you wouldn't have to write a book about admin to solve your worst admin problem. The market would offer a ready supply of consultants to help with these and other admin problems. Affordably.

*Everyone would have scaffolding to support their own private cathedrals in time.*

I've said that thinking about admin saved me. One concrete way involved my email. Thinking about admin forced me to find an architecture for escaping it. An architecture that, in my dreams, everyone would have.

I had tried for years to turn off my email, to leave it behind for stretches. To apply that clever consulting tip about doing things only once, I'd plan to check email a few times a day, only when sitting at the computer, so I could efficiently respond to new messages. This would work for a little while, usually right after I came back from a meditation retreat. Freed from email by the retreat's policies—which I would typically resist, resent, and then ultimately follow with relief—when I got back, I'd find it hard to remember why I had *ever even wanted* to read my email. I would return to life smugly uninterested in my inbox. But by day two, I would be back to checking my email dozens if not hundreds of times a day. (Yes, hundreds; if you glance at your phone fifteen times in an hour for fifteen-plus hours a day, you're at two hundred and twenty-five.)

When I began to fear the burdens of divorce admin waiting around every corner, I had to take action. I needed a way to protect myself—my mind, my heart, my time. Inspired by Abraham

Joshua Heschel's idea of the "sanctification of time" and the Jewish Sabbath as a "great cathedral" in the "architecture of time,"[63] I began to turn off my email on Friday evening.

I didn't grow up with any rituals for Shabbat, as I converted to Judaism as an adult. In recent years, my connection to Jewish ritual and my desperate need for a new way to manage admin have converged. My email Shabbat is not a strict observance of traditional Jewish law. Far from it. I can still be reached by text and phone sometimes, for instance. But something about the history and community of people observing this day as sacred, as set apart from the rest of the week, has supported me in carving out this freedom I had so rarely been able to claim.

I'm not alone in finding respite in a nontraditional relationship to the Jewish Sabbath. Tiffany Shlain initiated something similar in her extremely tech-savvy family (she's the founder of the Webby Awards and is married to a professor of robotics). She told NPR's Krista Tippett that the idea emerged from the family's experience of connection when her father was dying of brain cancer:

> When I would go over to see him for—sometimes it was just one good hour a day, we would turn off our phones and just be incredibly present with my dad. And when he died, I [said to my husband], "Can we as a family"—and we have children—"can we turn off the screens for Shabbat?" And we call them our "technology shabbats." . . . We're now just starting our seventh year of doing it. And it has been the most profound thing.[64]

In theory, anyone could do this—turn off the phone for a while; stop working and simplify life for a time—so why is collective action needed?

Perhaps for you, this is easy. For me, though, there was something necessary about having an external structure to support these efforts—something that went beyond myself. I have tried to turn off email at a certain hour of the evening on weekdays, too.

But I find that impossible to maintain. Without an architecture to lean on, without the sense of a broader community engaging in some version of this practice, I don't have a chance. Surely everyone shouldn't have to convert to Judaism to experience an email Shabbat. So what can be done?

Changing norms would be a start—being happy for people, rather than affronted, when they are away from their phones or email could help. But as long as our strategies are entirely individual, everyone has to adapt in each instance.

Some workplaces are trying out policies against sending nonurgent emails after a certain time of the evening or on weekends—or even automatically deleting emails sent to employees who are on vacation.[65] These innovations may help with work admin, but won't help most people with life admin. A few groups have declared national or international days of the year for putting down devices or turning off email.[66] These special days may raise awareness of a problem many of us face, but even if they caught on, a once-yearly event wouldn't support structural change to the rhythm of our lives.

A weekly, global, secular Shabbat—when everyone who wanted to opt out of email could participate and everyone else was well informed about this widespread practice—could give support to its users and alleviate the need to explain being off email for one day a week.

*We'd all know and could decide whether to follow Correspondence Rules for the Admin-Conscious.*

I've been trying to start a trend. To coin an acronym: NNR. No Need to Reply.

NNR aims to protect the recipient of an email or a text from any feeling that he should respond. So you can send along an alert or a reminder or a wish without encumbering the other person. As in this text: *Email invitation coming your way for dinner on the fifteenth. NNR.* Or *Reminder: Dad's birthday today. NNR.* Or *I'm completely drowning. Send prayers. NNR.*

NNR is just one example of how we might liberate one another in our correspondence. Saying less isn't the only way to be more admin-considerate. As in other areas of admin life, we have a choice: Either make admin take less time (improve *efficiency*), or transform admin into more of an end in itself so it feels less admin-like (add *value*).

Adding value is tricky. I could append a joke to every email I send out, but only some people will share my sense of humor. For the rest, the joke is, at best, a small time-suck; at worst, it makes them cringe and resent reading it. Ideas for adding value to other people's admin experience should be tried cautiously and carefully, if at all. With certain select communities I know well, I have experimented with adding a poem to a scheduling email. To be safe, I put it at the very end, so no one in a hurry has to wade through it to get to the business side of things—and I make clear there is no need to read. (I send poems to close friends sometimes too, especially in hard times, with a clear NNR at the end.)

Efficiency promoting ideas are likely to have broader uptake than attempts to add value, which require shared tastes. NNR is one example. Another proposal is for people to write any requests in the subject line or first line of any email. Yet another is for people to reserve texting for truly time-sensitive matters and put other requests in email.

This last item may sound good to some of you but not to others—highlighting the point that people's communication preferences vary. Lillian and Wendy told me adamantly that they prefer "just picking up the phone" to email these days; Leora and others told me just as adamantly that the phone is time-consuming or even draining and they have to carve out special windows to face phone interactions. Technology-preference mismatch is a major cause of admin friction.

In my utopia, information about our individual communication preferences would be transferred seamlessly to all comers. People would just *know*. In reality, with time, perhaps pop-up windows could give relevant information when someone types your email address or phone number—for instance, a notification that you are traveling, or that you have dispensed with email pleasantries, would appear before they finish composing or hit send—or an alert saying you prefer short texts would pop up just as someone types word number twenty-five in a text addressed to you. For now, perhaps we could start kindly letting each other know our communication preferences, asking when in doubt, and collectively agreeing not to take offense. (To get the ball rolling, I offer a reward—an Admin Coupon perhaps—to the first friend who admits a preference not to receive any more poems from me.)

*Schools would teach Admin Ed.*

Imagine a world where you were taught Admin Ed in school—and everyone else was too. Think how much more capably we would face admin throughout our lives if, from a young age, we had the tools to do this work well, when necessary, and otherwise to see it, name it, and challenge it.

Admin is associated with adulthood, but some kids have a lot of admin to do already. This is most obvious for kids living in poverty. Andres Gonzalez, whom I heard speak about the Holistic Life Foundation—which he co-founded to teach yoga, meditation, and much more, to kids in underserved communities—shared with me his sense of the admin difficulties faced by the kids he works with.[67] These include homework management (not just doing it but remembering to take it to school and turn it in), permission-slip management (getting signatures and turning in), keeping track of belongings (locating and not losing backpacks, clothes, and more), and self-care (like remembering to brush teeth and hair).[68] The items on his list present challenges for many privileged kids as well.

How does anyone learn this stuff? Some people surely learn it from their parents. But why assume parents are good at these things? Either by personality or by virtue of the admin overwhelm (and other challenges) in their own lives, parents may be avoiding or denying or just failing. We don't assume parents will teach math. We shouldn't assume parents will teach admin.

In my dreams, every school would teach Admin Ed. The starting point would be executive-function admin—developing tools for organizing and remembering what we need to do. We'd learn actual office-work skills, both clerical and managerial. We'd gain technical skills like scanning, faxing, filling out forms, using spreadsheets, and setting up (and sharing) electronic calendars, for starters. And we'd acquire financial literacy, from efficient ways to budget and pay bills online, to strategies for investing and long-term planning.[69]

On the managerial side, Admin Ed would teach techniques for delegating effectively and for using an "affiliative style" to influence others. And we'd develop skills in listening and in articulating our own needs. (Imagine if from a young age we all mastered the art of taking turns talking, listening, and reflecting back—as in the Listening Game described in chapter 11; think how different many of our conversations about household labor, and much else, would be.[70]) We'd learn about stickiness, so we'd appreciate the workflow pattern of admin *before* its initial landing points in our lives. We would also learn to manage our own relationship to technology—something so many of us struggle with—so that we control technology rather than technology controlling us.[71]

Admin Ed would help us all become more socially conscious around admin. We'd learn to minimize admin impositions on others—and learn to notice (and express appreciation) when others help with admin. We'd become aware of admin's varied burdens across the life cycle—around both happy and difficult events—and across different identities and degrees of privilege. This would enable us to make conscious choices around big moments rather than being surprised by admin onslaughts—and to replace admin judgments with admin compassion.

The culmination of Admin Ed would be an experiential "lab" in which students conduct Admin Audits of their school and other local institutions (like hospitals and libraries). Working together in teams, students would assess the admin these entities inflict—for example, back-to-school forms—and propose improvements. Schools and other institutions would then have the opportunity to model good neighborly admin behavior going forward.

If we want more than isolated improvements in our individual lives, we need awareness and ingenuity around admin. And first and foremost, we need visibility. We need a society where everyone sees admin, so we can work together to lessen its collective drain on our time and our energy.

Admin Ed can help the next generation. For now, I hope the ideas in this chapter inspire you to dream big and imagine a different world. The admin revolution needs you.

# Epilogue

*Walk around feeling like a leaf.*
*Know you could tumble any second.*
*Then decide what to do with your time.*

—Naomi Shihab Nye, "The Art of Disappearing"

These days, I regularly win back time from admin—and appreciate it when I do. I keep my eyes peeled for moments when I can resist. With admin that needs doing, I try out an adminimalist strategy first when possible, as I did when searching for a new babysitter recently. I emailed a few former babysitters who I hoped would have contacts with qualified people. When one emailed back to say she'd found me someone, without my having to post my ad and launch a big search, I celebrated the gift of Admin Savings Time by planning a leisure activity I'd been wanting to try (something called "floatation therapy").

I also create windows for getting necessary admin done. I aim to schedule a study hall with a friend or colleague at least once a week—either to do admin or to exclude admin in favor of other priorities. Some weeks, scheduling a study hall is itself too much admin. When I want some extra structure without the hassle of coordination, I use the chart I made for a solo study hall.

My systems are not optimal, but they are working pretty well.

The endless recurring to-do lists in my calendar are largely a thing of the past. For the moment, I have also stopped spending my time trying out to-do-list apps. Instead, I rely on the Notes app on my iPhone for my long list of tasks, and on especially busy days, I write a paper to-do list. Each day's page begins with a box of several important-but-not-urgent items I aspire to do every day: meditate, exercise/yoga, write, pray. I have an actual physical clipboard labeled "Current," where I put snail-mail items that need to get dealt with, and a clipboard labeled "Tax," where I save donation receipts and other documents for tax time. In lieu of filing, I have a box for any papers I might need from this year (for example, bills paid with date notation, backup originals of a few essential items I've scanned). I'm still an Avoider with my snail mail, which can pile up, but at least I have places to put things when I do choose to sort. Nearly all these innovations were inspired by particular interviewees.

I try to see admin before it lands and make quick decisions about how to respond. I text or email recommendations to people the moment they ask, while they're still standing there to dictate their email addresses. If someone requests something more complicated, I ask the person to email me the relevant information in a "clean thread" (an email with no history of past exchanges) that I can forward along to the desired contact. I take photos of information, and rather than scanning, whenever possible I photograph permission slips and other documents that need signing and emailing. When I'm blessed by an admin near-miss, I try to savor the feeling. *No lost-luggage admin*, I'll think over and over while leaving the airport with my bags.

When a big admin event is coming up, like a lengthy and demanding divorce meeting, I've gotten better at trusting my own methods. I schedule a pre-meeting with my financial adviser, to go over anything concrete, and a few minutes with my lawyer right before the meeting. I jot down notes as they occur to me. And I trust that whatever thinking and gathering needs to happen will happen in the day or two before. Admin for the big meeting doesn't take over the whole two weeks prior. But I also acknowledge that it will take some time and that it's real.

The market has also given me a few admin bright spots. I discovered an app that handles divorce finances; you and your ex can each log shared expenses and upload receipts (for the cost of afterschool activities or medical bills, for instance), and the app will charge you or your ex the right percentage (according to your formula) and keep a record of it all for you both. The company's founding principle is Don't Ever Talk to Your Ex About Money Again. Brilliant. And to my delight, my employer fixed my pretax-dollar Flexible Spending Account problem by consolidating service of that with my health-insurance company. Now that company will *automatically* feed the amount that insurance doesn't cover right through my pretax-dollar account and *automatically* deposit the remainder in my bank account. These developments underscore to me the huge difference markets and employers can make in our individual admin burdens.

For happier events, like planning my daughter's birthday party, I indulge the parts I like and ask myself what's worth doing. When the options overwhelmed me before party time this past year, I finally sat down and made a quick chart of my values, including things like her getting to invite whomever she wants and getting to play the way she wants. Immediately, this helped me rule out some glossy options, such as Robot Building, which charges by the kid and structures every minute. Making a chart of what matters in this endeavor is not something I would have thought I had time for, but it ended up saving precious minutes and mental energy (and money).

I continue to turn off email from Friday night to Saturday night for my email Shabbat. I scurry around frantically before this hiatus starts, trying to get everything possible done; that's one stressful hour on Friday early evening. Sometimes it feels like I'm literally wrenching myself away from the to-do list to light Shabbat candles. And then, not long after, it's like I remember something I'd forgotten. Something quieter, more joyful. My kids look forward to and comment on this change in my energy. By Saturday night, I feel light-years from the difficulty I faced shutting down on Friday. I usually don't want to open my email, and, when I can, I wait until Sunday.

Admin certainly comes in and torments me sometimes. At other times, though, I've come to feel it as a relief. It can even be a siren song, threatening to lure me away from my important-but-not-urgent endeavors like writing. Amidst all, it feels like a miracle to be finishing this book. I've found that writing your first book—or at least, writing *my* first book—is a lot like so many other firsts. The first time you go skiing, for instance, or your first kiss. Even if you know you're doing something monumentally exciting, the sheer quantity of fear may make it hard to enjoy. On the way to my first junior-high dance, my two friends and I sat in the back seat of my mother's car, petrified. Rigid with fear. My mom said gently, "It'll never be this exciting again."

As I emerge from this journey, my admin life has changed in fundamental ways. Some areas of admin have been made smaller; some, more manageable. And some admin feels very different than it used to. Making admin visible, counting it as work, and reflecting on how I feel while doing it has made the admin that I know needs doing feel better. But some of the change is the result of the particular circumstances of my life—specifically, my divorce. Though I began this project largely as a Reluctant Doer, I have always had my strongest streak of Super Doer when it came to my kids. This was especially true for projects I found interesting, like searching for the right school. These days, even the dullest admin for my kids—like getting physical forms completed and submitted for summer camp, say—feels pretty okay to do. I think the main reason is that it feels like work I do directly for my kids, whom I miss half the week. And so in some way the kid admin connects me to them rather than getting caught up in unrealized hopes or complications of the adult world.

Admin is woven into the very particular details and values of our lives, and so the way we experience it is intimately related to those details and values. For you, so it is for me.

Two days before the party for my daughter's seventh birthday, I was in the throes of trying to finish a first draft of this book amid the admin hell that was my life. Shortly before dinner, I sat at my computer hoping to make good use of thirty minutes of writing time.

My daughter came running in to the room and announced, with a big smile, "Mom, I made some surprises for you that I'll give you on my birthday!"

"For me, on *your* birthday? Why?"

"Because I love you. And also because you helped with the party."

All that admin doing, or even some of it, was visible. For me, in that moment, it was enough.

## Note to the Reader

# Inviting Your Ideas

Dear Reader,

I've learned a tremendous amount from people who've shared their admin stories and strategies with me. Some of my favorite *Ideas to Try* are collected in Appendix A, and I'd also love to keep learning how we can make things better. I'd like to hear your ideas.

Do you have admin strategies that have worked for you personally or in your relationships? Do you have insights into how our society could improve around admin—for markets, employers, insurers, airlines, schools, government, or social norms? Or do you have admin stories to share, including about how you've spent your Admin Savings Time?

I invite you to send your admin insights and stories to me at Ideas@LifeAdmintheBook.com. By sending me your ideas, you are giving me permission to use them in the things that I write and say, but I promise not to use your name unless you give me express permission to do so. And I'll try to find a way to share the best strategies and ideas with other interested readers.

Thanks so much for your time, and I look forward to learning from you—

Liz Emens
Columbia University

# Acknowledgments

*My apologies to everything that I can't be everywhere at once.*
*My apologies to everyone that I can't be each woman and each man.*
*I know I won't be justified as long as I live,*
*since I myself stand in my own way.*
*Don't bear me ill will, speech, that I borrow weighty words,*
*then labor heavily so that they may seem light.*

—Wisława Szymborska, "Under One Small Star"

My deepest appreciation goes to over one hundred people I cannot acknowledge by name, but you know who you are. You are my interviewees and brainstorming-session participants. I am so grateful for your candor and your insights. And my hearty thanks also to the many more friends, colleagues, and acquaintances who have informally shared their admin ideas and experiences with me.

Having always written for academics before, I wrote an article version first, containing 324 footnotes. And the first draft of this book was much longer and had more involved endnotes, with longer source lists and further explanation. I was persuaded to cut many of those citations. A fraction of them can be found in my "Admin" article, though many other sources I read after writing that. (If you're interested in the article, the citation is included in the endnotes.[1]) My profound gratitude goes to all those from whom I have learned, in person or through their printed words, while writing this book.

There are also many other people I would like to acknowledge by name but cannot, not for reasons of anonymity, but simply because I do not have enough pages. You have shared ideas or insights or people to contact; you have done research or suggested research; you have read pages or chapters or passages or the whole manuscript; you have tolerated my absences or my distraction; you have supported me directly or indirectly; you have contributed to this book in more ways than I can say. A tiny fraction of you are named here, in addition to those named in the acknowledgments to my earlier article: Rachel Adams, Adefisayo Adetayo, Erez Aloni, Jill Anderson, Susan Appleton, Ellis Avery, Chad Baker, Sujatha Baliga, Annie Barry, David Beizer, Noa Ben-Asher, Laura Brady, Abigail Bray, Samuel Bray, Heidi Brooks, Vincent Brunsch, Nathan Bu, Jimmy Bui, Jessica Bulman-Pozen, Debora Cahn, Iliria Camaj, Mary Anne Case, Alicia Cho, Andrea Clay, Mathilde Cohen, Jess Cole, Susanna Cole, Ruth Cordero, Anne Coxe, William Cranch, Allison Daminger, Odalys Diaz, Amy DiBona, Justine di Giovanni, Brett Dignam, Antonia Domingo, Rashmi Dyal-Chand, Jason Edwards, Avlana Eisenberg, Niva Elkin-Koren, Sophie Elsner, J. Richard Emens, Alaine Emens, Jack Emens, Carol Emens, Ariel Emens-Asher, Micah Emens-Asher, Carol Esler, Jeffrey Fagan, Lindsay Farmer, Robert Ferguson, Timothy Fisher, Rina Fujii, Kris Gaziano, Jacob Gersen, Jeannie Suk Gersen, Haben Girma, Maeve Glass, Suzanne Goldberg, Andres Gonzalez, Logan Gowdey, Michael Graetz, Timothy Gray, Kent Greenawalt, Jamal Greene, C'naan Hamburger, Philip Hamburger, Donna Harati, Bernard Harcourt, Jill Hasday, Michael Heller, Rabbi Lauren Grabelle Herrmann, Bert Huang, Clare Huntington, Mary Emens Ihle, Paul Ingram, Olatunde Johnson, Kate Judge, Karen Kadish, Adam Katz, Abbey Keister, Jody Kraus, Anna Krauthamer, Michael Lanci, Laura Lane-Steele, Sarah Lawsky, Juhyung Harold Lee,

Georgia Lee, Carol Leibman, Gillian Lester, Dan Li, Jed Lippard, Mika Madgavkar, Florencia Marotta-Wurgler, Serena Mayeri, Brett Mead, Andrea Metz, Gillian Metzger, Alison Miller, Anna G. R. Miller, Anne Miller, Carson Miller, Ian Miller, Leigh Miller, Rick Miller, Henry Monaghan, Mary F. Moyer, Sendhil Mullainathan, Nancy Murray, Dr. Home Nguyen, Yerin Pak, Bruce Pettig, Christina Duffy Ponsa-Kraus, Taylor Poor, Leonard Powell, David Pozen, Alex Raskolnikov, Theodora Raymond-Sidel, Anthea Roberts, Russell Robinson, Tracy Roe, Matthew Rose, Clifford Rosky, Christine Rua, Kelsey Austin Ruescher-Enkeboll, Chuck Sabel, Leigh Anne St. Charles-O'Brien, Sharon Salzberg, Adam Samaha, Carol Sanger, Casey Santiago, Barbara Schatz, Davida Schiff, Elizabeth Scott, Robert Scott, Rena Seltzer, Susanna Shannon, Matt Shapiro, Ellen Sharabi, Robin Shulman, Michelle Smyth, Annie Stein, Bernardo Stein, Ilan Stein, Ryann Wahl Stevenson, Timothy Stewart-Winter, Mark Stopfer, Susan Sturm, Noella Sudbury, Wendy Suh, Lauren Sun, Cass Sunstein, Thomas Swanson, Eric Talley, Kristen Underhill, Caroline Voldstad, Joshua Wan, Huvie Weinrich, J.C. White, Joan Williams, Patricia Williams, Kate Witchger, Marty Witt, Bea Wolper, Timothy Wu, Rebecca Yergin, and Kenji Yoshino, as well as participants in workshops held in various universities, which are acknowledged by name in my "Admin" article. A special thanks goes to the research assistants who helped me organize and lead brainstorming sessions and interviews that contributed so much to the book. My thanks also goes to the Peter Drut and the Philippe P. Dauman Faculty Research Funds and the staff of the libraries at Columbia Law School and Harvard Law School, as well as all of the administrative support staff at both of those institutions.

My gratitude also goes to my careful and insightful—and patient—editors Bruce Nichols and Daniel Crewe and their talented teams at Houghton Mifflin Harcourt and Viking Penguin Random House UK. I am especially appreciative of the support and care of my agent, Sarah Chalfant, as well as Rebecca Nagel and all those who work with them at the Wylie Agency.

To the rest of you who contributed but go unnamed, please accept my silent recognition.

To anyone who has found the time to read this book when your time is so precious and cannot truly be repaid, I am grateful. May we change the world, with grand gestures and structural interventions, yes, but also with one disruption of one imposition at a time.

Appendix A

# Ideas to Try

*Life Admin* contains many ideas I've gathered on my admin travels. To give you a jump start on trying a few, and to spare you a little list-making admin, I've collected some of my favorites, organized into three categories:

- The "Urgent List" is for when you need immediate relief.
- The "Hacks List" is for when you are prepared to make a system improvement.
- The "Love and Relationships List" is for when you want to make life better alongside your partner, your friends, your family.

## My Favorite Ideas to Try—the Urgent List:

### *For Immediate Use When You're in an Admin Onslaught*

1. **Forget the search for a magic tool and embrace a simple to-do list.** Use no (or few) categories, on actual paper, or if you prefer a digital version, try the Notes app in your phone.
2. **Start bypassing the to-do list when you face real-time admin requests.** Email or text someone the information she

wants while she's still standing there—so it never goes on your to-do list.

3. **Take photos to keep and transfer information.** When asked to fill out a form and sign it and scan it, try sending a photo of the signed form instead. Need to copy a recipe or remember where you parked your car? Take a photo with your phone.

4. **Piggyback on others' efforts, experience, and wisdom.** If you've been hit with some awful admin, find someone who's been through it and ask practical questions: *What person or site or strategy helped you the most? What do you wish you'd known or done differently? What's the very first thing I should do?* The person may be happy to bestow on you all that knowledge that she never wanted in her mind in the first place—and help a friend in the process.

5. **Try an Admin Purchase.** If you can afford it, buy your way out of some admin. For instance, order the cake delivered.

6. **Try an Admin Study Hall.** Get some company for your admin. And don't forget the rewards.

7. **Request your own admin babysitter.** Contact a friend or family member and ask if they can help by emailing or texting you at certain intervals. E.g., "Tell me to open the hospital bills at 8:00. Email at 8:30 to ask if I've finished. If not, email me again after a 30-minute extension. After an hour, send congrats and encourage a reward, whether or not I finish."

8. **Use an "affiliative" style with Customer Service.** Ask for the name of your interlocutor, write it down, and ask if you can follow up with him. Treat him kindly, like he's a human being (since he is) and like you're on the same team. It may make him want to help you, and knowing his name may help you later.

9. **Try connecting to the meaning.** When you have to do the awful admin, ask yourself what goal or value or opportunity it serves. (Research suggests you'll benefit if you can feel even a few moments of genuine gratitude each day.)

10. **Spend your Admin Savings Time well.** If you save yourself an hour, spend that hour doing something you really want—or need—for yourself.

## My Favorite Ideas to Try—the Hacks List:

### *For a Moment When You Are Ready to Make a System Improvement*

1. **Take an inventory of your murky admin.** Pick out the items that have lingered on your to-do list, and try to identify what's making them hard. Then tackle them with a reward or with the help of a family member or friend (in a "murky admin consult").

2. **Divert some admin away from you (permanently).** For instance, create a spam email account and give out that address for retail purchases and tickets. Then open that account only when you need to search for something you ordered.

3. **Start an Admin Documents Repository.** Scan IDs and other important documents and keep them in a folder you can access digitally at all times (on Dropbox, for instance).

4. **Hack your biggest admin process failure.** What's your least efficient admin process? (Mine involved putting my to-do list in my calendar as a recurring entry.) Figure out why the process keeps happening this way, face what's needed to solve it, and carve out the time to solve this problem for the benefit of your future self.

5. **Find ways to make things end.** For instance, try writing No Need to Reply (NNR) on texts and emails. Save others time; they might even return the favor.

6. **Use the Admin Pleasures Inventory to make a system improvement.** Mark where you fall on each item, and try to identify one way your current approach to admin doing could be reshaped to better fit your particular profile. For instance, if you care about aesthetics, but have no time to make your admin beautiful, try buying one item—like a special pen or notebook—to use when tackling admin.

7. **Outsource the administration of the task as well as the task.** If you get help with childcare or household chores, paid or unpaid, try asking for help with the admin of those tasks. For instance, if someone helps you with cleaning, see if the person might take charge of restocking any necessary supplies (and getting reimbursed). If a family member helps you with childcare, empower the person to decide what the kids eat—or even to plan meals and shop with the kids.

8. **Try out a different admin personality—and use it to try new strategies.** After you determine your admin personality (the Admin Personalities Quiz in Appendix B may help), embrace a different personality for a week and see what changes work for you.

9. **Make default plans.** For exercise and vital leisure plans, book something that recurs in your calendar, with friends or via classes. If it's with friends, make a rule that whoever cancels has to take charge of rescheduling.

10. **Top your to-do list with your good-day short list:** Make a short list of (3–5) important-but-not-urgent things you aspire to do each day, to make the day a good day, and put those at the top of your to-do list (always).

## My Favorite Ideas to Try – Love and Relationships:

### *For Moments When You Want to Make Life Better in Relation to Others*

1. **Try low-admin socializing.** You can do this for kids (with a group playdate) or adults (with coffee or drinks or a movie). How to do it: *Do not* email/text everyone asking for replies. *Do* email/text everyone to say "I'll be at the playground/café/theater tomorrow at 3:00. Any and all are welcome to join."

2. **Unstick yourself: create systems for seamless transfer of information.** For instance, set up shared calendars. (Many people like electronic versions; if you prefer paper, taking a photo of the calendar lets you have it everywhere and send it to others.) Start a list of all services you need to call—electrician, cable provider, babysitter—and put it someplace easily accessible by everyone. (If it's too much admin to compile the backlog, start fresh with anything new that comes in.)

3. **Play the Listening Game around admin.** You have 3 minutes to say what admin is on your mind (or what admin you hate or secretly enjoy). Your partner has 2 minutes to reflect it back to you. Reverse positions and repeat.

4. **Recognize that most household projects involve 3 key parts: planning/research, decision-making, and execution.** If you want to share a project, decide who does each part. For instance, one person can do the front-end research, and the other can do the back-end implementation. You can make the decisions together or make one person the decider.

5. **Find ways for you each to illuminate your admin contributions.** Put a list of admin tasks somewhere visible, to you

and your household, so everyone can check them off as they complete tasks. Or take a moment at dinner to ask each other what tedious admin you've done for the household this week.



6. **Ask the Admin Question before giving a gift.** Does the gift force the receiver to do any admin? Can you do it for her, as part of the gift (if you're up to it), or should you give her something else?

7. **Give the gift of your admin-doing.** Can you offer to do some admin for someone you love? Especially in times of crisis, if you can take over an admin task (Call a clinic to ask for an earlier appointment? Research funeral homes?), you may be giving a unique gift when it's most needed. The fun version of this can be an Admin Coupon—for the person to use as they like.

8. **Watch people light up when you notice and really thank them for their admin doing.** If someone else does admin for you, or saves you some admin time, say thanks for exactly what they did and why it was awesome. (They may repeat the performance!)

9. **If you or your friends are single, put a premium on prospects who are good at admin.** A person's being good at admin may contribute to their *partner's* happiness and success.

10. **Help kids see and learn to do their own admin.** For example, start to praise kids for remembering as much as for doing a chore, issue reminders only when needed, and do so in a way that builds kids' capacity to remember next time. (As in, "What are you going to do first after getting up from dinner?" rather than "Remember to clear your plate.")

Appendix B

# Admin Personalities Quiz

## What's Your Admin Personality?

Below is a series of questions about how you relate to life admin —the office work of life. Choose the answer that is closest to how you act or feel, even if none of the answers fits you exactly. *In fact, the answers probably will not fit you exactly, for reasons discussed at the end of the quiz.* Some of the questions are hypothetical, so they may not apply to you right now, but imagine you are dropped into the situation. What would you do?

There are no right or wrong answers. And no admin personality is best. The aim of the quiz is to understand oneself better in relation to life admin.

If you have read the book, you may recognize the personalities in the answers. And if you haven't read the book yet, the quiz should begin to acquaint you with the personalities that are developed in the preceding pages, particularly in Chapter 3, where they are introduced, and Chapter 12, where we draw strategies from each personality.

1. How do you keep track of your to-do items for your life and home?

   *Note that this and all questions refer to office-type work in your personal life, not in your job.*

   A. I have a method that works really well for me, so I usually feel on top of things.

   B. I don't have time to create a really good system for keeping track of it all, but my methods work well enough to make sure the essential things get done.

   C. My partner nudges me, sometimes with a list, sometimes with a pleading text or email.

   D. Why would I need a list? There's not that much to do.

2. How do you deal with your snail mail at home?

   A. I collect it every day and deal with it promptly.

   B. I usually collect it and pull out the important stuff. I make myself deal with the important stuff by the due dates.

   C. My mail pile topples over unless someone else takes care of it—or until something overdue catches up with me.

   D. Nothing important seems to happen in snail mail, though I don't really check.

3. Imagine that you and your immediate family need to plan a party for a close friend. Which statement is most likely to be true for you?

   A. I'm great at throwing parties and enjoy doing it well, whether I do all the work myself or delegate really efficiently.

   B. I know the party will end up on my plate, and I'll do it, but at the cost of much time and energy.

   C. It's fine with me for the party to take place, and I can be part of the initial discussion if they want, but I probably won't be that much help, practically speaking.

   D. Just tell me when and where to show up.

4. How do you relate to *life admin*?

   *The term* life admin *or* admin*—in other words, the office work of life—includes managerial tasks, like making decisions and delegating, and secretarial tasks, like scheduling, filling out forms, and paying bills.*

   A. I mostly get admin done and feel pretty good about it.

   B. I mostly get done whatever admin needs to get done, but I wish I didn't have to.

   C. I try to avoid it, but I don't feel so good about that.

   D. I mostly stay away from the thing you're calling admin (if it's a thing at all).

5. You receive an email that was sent to multiple people (including your partner, if you have one). The email makes a request for information in the second paragraph. Which answer best describes what happens next?

   A. I am likely to respond first and be as helpful as I can (quickly) be.

   B. I am often the one to respond, not because I want to, but because I doubt anyone else will.

   C. Someone else is likely to respond before me.

   D. I'm not likely to read as far as the second paragraph.

6. Which of the following statements sounds most like you in relation to household bills?

   A. I have a good method for paying my bills, so they never weigh on me.

   B. I pay my bills on time, but I wish I didn't have to deal with them.

   C. My bills pile up, and eventually guilt or late fees force me to deal with them—or someone else gets to them before I do.

   D. I don't seem to pay many bills.

7. How do you and your partner (or housemate) divide up your shared admin?

    Shared admin *refers to the office-type work of managing a shared household. If you don't have a partner or housemate, imagine the most likely outcome if you did.*

    A. I am in charge of most of it. I may delegate some specific tasks, but I make sure everything gets done to my satisfaction.

    B. We aim to divide things up evenly. If someone does more, though, it's probably (or definitely) me.

    C. My partner/housemate is in charge of more of it. Some items are delegated to me, but I'm not very quick, so those items sometimes get done without me.

    D. It doesn't seem like there's much shared admin to do.

8. What metaphor most closely resembles your feeling about life admin?

    A. It's like the air I breathe, and it comes and goes pretty smoothly and constantly.

    B. It's like a wet blanket weighing on me—something heavy and relentless.

    C. It's like a mosquito in the room, annoying but mostly avoidable.

    D. No metaphors come to mind, since I don't really see admin as a thing.

9. How do you feel about talking about life admin?

    A. There's mostly no need, since I've got it under control, but I can talk about it if necessary—for instance, if it involves joint planning or delegation, or if you want advice.

    B. It would be a relief to vent some frustration about doing admin, but who has the time when there's so much to do?

    C. Please let's not talk about this stuff; I'm trying to get away from it.

    D. What is there to talk about?

10. Is there anyone in your world who you think does *less* life admin than they should?
    A. Most people.
    B. The person who leaves it to me (for example, my partner, housemate, or family member).
    C. Probably me.
    D. I don't think so.

11. How do you think other people perceive you in relation to life admin?
    A. As the go-to person for getting things done well and fast (perhaps even effortlessly, or so it appears).
    B. As the default contact person and admin Doer in many situations, even when I'd rather not.
    C. Not the best person to contact for life admin, though I may be in the loop.
    D. Mostly unaware of the admin, if any, that may be going on.

12. Do you ever suffer from madmin mind?

    *(*Madmin mind *is a mind spinning with so much life admin that it feels like it's in overdrive.)*

    A. Occasionally, but mostly I stay on top of things so they don't spin out of control.
    B. Unfortunately, yes. Too often.
    C. I've seen madmin up close, mostly in others, and it's pretty scary—it's one reason I try to avoid admin.
    D. No. I'm not sure what could make that happen to someone.

13. You're finally in bed after a long day and have just turned off the lights. Which best describes you at this point in relation to life admin?
    A. You make a mental checklist of your tasks for the next day, feel satisfied about the prospect of another productive day, and promptly fall asleep.
    B. You toss and turn at the thought of all the things you need to

do over the next few days (or weeks or months), or you fall asleep promptly but wake up at four a.m. with your to-do list in mind.

C. You suddenly remember that you need to pay a long overdue bill but feel so tired and settled in that you fall asleep.

D. What admin? You stretch, appreciate how comfortable your bed is, and fall asleep immediately.

*Scoring Instructions*

It's possible you noticed a pattern to the answers: Each letter always corresponds to the same admin personality. (This makes scoring your result very low-admin!)

For each question, simply count how many times you chose each letter (*a* for Super Doer, *b* for Reluctant Doer, *c* for Admin Avoider, *d* for Admin Denier). Or, for even less admin, just glance at what answers you circled most often to get a sense of your pattern.

**HOW MANY OF EACH LETTER DID YOU ANSWER?**

| | Super Doer<br>A | Reluctant Doer<br>B | Admin Avoider<br>C | Admin Denier<br>D |
|---|---|---|---|---|
| Total | | | | |

## Interpreting Your Score

The column with the most answers is your primary admin personality. (See below for a primer or a refresher on the personalities.)

The column with the second most answers is your secondary admin personality. As discussed in Chapter 3, our admin personalities can vary across relationships and across contexts. For instance, you may have one personality in relation to a particular partner or housemate and another personality in relation to extended family. Or you may have one personality in relation to planning parties and another in relation to snail mail.

So, for instance, if you circled eight *b*'s (the Reluctant Doer column) and five *c*'s (the Admin Avoider column), then your primary personality is Reluctant Doer and your secondary personality is Avoider.

The closer the scores between your primary admin personality and any other columns, the greater your variability. So if you had six in the Super Doer column and seven in the Reluctant Doer column, then it's likely that your admin personality differs across relationships or contexts.

## The Personalities: A Refresher

The personalities capture a combination of action and feeling. This chart sums them up:

| | **Feeling Good** | **Feeling Bad** |
|---|---|---|
| **Doing** | Super Doer | Reluctant Doer |
| **Not Doing** | Admin Denier | Admin Avoider |

Those in the top row—Doers—are mostly getting admin done. Those in the bottom row—Non-Doers—are mostly not getting it done. Those in the left column feel pretty good about their admin, and those in the right column do not.

These are archetypes. Most of us are really hybrids. And our answers can also be affected by the sheer quantity—large or small—of admin in our lives and households right now. But these are the basic four (clockwise, starting in the top left corner):

*Super Doer:* You are getting the admin done—by doing it yourself or explicitly delegating it to others—and you feel basically on top of it and therefore pretty good.
*Reluctant Doer:* You take care of the admin that needs doing, but you wish you didn't have to.
*Admin Avoider:* You avoid admin as much as possible and feel not good about it, either because the consequences of your non-doing catch up with you or because you feel bad that other people are doing it for you.
*Admin Denier:* You generally don't see admin as a problem, or even as a thing; you do very little of it and don't feel bad about that.

## Note to the Reader on How to Use This Quiz

The idea of these admin personalities grew out of my thinking over several years, but the quiz has not been tested empirically. I offer it as a heuristic device, an interpretive tool, for thinking about admin and one's own relationship to it. Note that each personality can offer insights and strategies into how we might choose to address admin (as discussed in Chapter 12).

The quiz may be interesting to take again with someone close to you in mind. Try to guess their answers, and then, if it's comfortable (or even enjoyable), ask the person to take it also so you can compare your guesses with their answers. Perhaps there will be something surprising to discuss.

# Notes

## Introduction

1. Other versions of the phrasing of this first question included "What kinds of admin, if any, take up time in your life?"
2. Arlie Hochschild and Anne Machung, *The Second Shift: Working Families and the Revolution at Home* (New York: Viking, 1989), 4. Their incisive analysis of the second shift, which includes some key parts of what I call life admin, has been an important catalyst in the study of work-family balance and household distribution of labor.
3. Some of the terrific books with insights into how overwhelmed we feel and how we might make things better include David Allen, *Getting Things Done: The Art of Stress-Free Productivity* (New York: Penguin, 2001); Tiffany Dufu, *Drop the Ball: Achieving More by Doing Less* (New York: Flatiron Books, 2017); Jancee Dunn, *How Not to Hate Your Husband After Kids* (Boston: Little, Brown, 2017); Daniel Goleman, *Focus: The Hidden Driver of Excellence* (New York: Harper, 2013); Craig Lambert, *Shadow Work: The Unpaid, Unseen Jobs That Fill Your Day* (Berkeley, CA: Counterpoint, 2015); Sendhil Mullainathan and Eldar Shafir, *Scarcity: Why Having Too Little Means So Much* (New York: Henry Holt, 2013); Marilyn Paul, *It's Hard to Make a Difference When You Can't Find Your Keys: The Seven-Step Path to Becoming Truly Organized* (Penguin, 2003); Brigid Schulte, *Overwhelmed: Work, Love, and Play When No One Has the Time* (New York: Farrar, Straus and Giroux, 2014).
4. Much of what appears in the book was written around the time when it happened or I learned about it, so the stories—my own and other people's—do not follow any neat timeline, nor do they purport to represent the current state of things.

5. The stories and insights in this book came to me in a variety of ways. Much of it was drawn from my own experience and reflection. Many friends, acquaintances, and family members shared their stories and strategies over a number of years. As an academic, I spent several years researching published sources on related topics, and I gave presentations at which audience members made many insightful points, frequently through personal narratives. In addition, I conducted interviews and brainstorming sessions with more than a hundred individuals. I found them largely through a process called snowball sampling in which one interviewee led me to the next, based on that person's sense of who would be most interesting for me to interview. To protect the privacy of all those who were generous with their time and their candor, I have changed their names and described them in ways that preserve their anonymity.
6. This book takes as its principal subject admin in the United States, even while incorporating some cross-national examples, and so the collective solutions are largely directed at US law and policy. The small part of my research that involved participants from outside the United States—for instance, the brainstorming session I conducted by videoconference with participants in Glasgow—as well as my discussions of admin when presenting this work overseas, suggested to me admin's global relevance as well as its local textures.

## 1. What Is Admin?

1. Judith Scott-Clayton, "Simplifying and Modernizing Pell Grants to Maximize Efficiency and Impact," 5, https://www.urban.org/sites/default/files/publication/93301/simplifying-and-modernizing-pell-grants_2.pdf.
2. Eric P. Bettinger et al., "The Role of Application Assistance and Information in College Decisions," *Quarterly Journal of Economics* 127, no. 3 (July 2012): 1207–8.
3. Benjamin J. Keys, Devin G. Pope, and Jaren C. Pope, "Failure to Refinance," NBER working paper no. 20401, National Bureau of Economic Research, Cambridge, MA, August 2014, 18–19, http://www.nber.org/papers/w20401.pdf.
4. Names have been changed throughout the book to protect anonymity, as explained more fully in Introduction note 5. For ease of exposition I sometimes avoid quotation marks, and I omit vocalized pauses and fillers (such as *um*, *er*, *like*, and *you know*) and repeated words.

5. Claudia Dreifus, "Seeking Autism's Biochemical Roots," *New York Times*, March 24, 2014, http://www.nytimes.com/2014/03/25/science/seeking-autisms-biochemical-roots.html. (Emphasis mine.)
6. Three countries other than the United States that do have some form of Medical Savings Account system are China, Singapore, and South Africa. See Justine Hsu, "Medical Savings Accounts: What Is at Risk?" World Health Report Background Paper 17 (2010), p. 5, http://www.who.int/healthsystems/topics/financing/healthreport/MSAsNo17FINAL.pdf.
7. See, for example, Nir Eyal, Paul L. Romain, and Christopher Robertson, "Can Rationing Through Inconvenience Be Ethical?," *Hastings Center Report* 48, no. 1 (January/February 2018): 10–22 (discussing this, among other arguments). There is also the issue of fraud. These steps offer documentation of the relevant expenditures, but there are other ways the employer could get this information, particularly for health-related spending that has gone through insurance first.
8. This is one reason why life admin is a broader category than household management, a distinction that will be discussed further in Chapter 4.
9. On the term "mental load," see, for example, Diane Ehrensaft, "When Women and Men Mother," in *Mothering: Essays in Feminist Theory*, ed. Joyce Trebilcot (Savage, MD: Rowman and Littlefield, 1983), 53. The popular comic about the mental load by Emma—originally "Fallait Demander" and then translated into English as "You Should Have Asked"—will soon appear, with other stories, in book form. See https://english.emmaclit.com/2017/05/20/you-shouldve-asked/; Emma, *The Mental Load: A Feminist Comic* (Seven Stories Press, 2018).
10. People apparently use the term *admin* to refer to the admin of life—not just of work—more often in the UK, for example, than in the US.
11. "Make a Plan," Department of Homeland Security, https://www.ready.gov/make-a-plan (the "Create Your Family Emergency Communication Plan" form is available at this site and, as of April 10, 2018, still spoke of the "PrepareAthon").
12. Ian Ayres and Peter Siegelman, "Race and Gender Discrimination in Bargaining for a New Car," *American Economic Review* 85, no. 3 (June 1995): 304–21; Mary Anne Case, "Developing a Taste for Not Being Discriminated Against," *Stanford Law Review* 55 (June 2003): 2273.
13. One participant, who'd studied law, termed it "obviating a disbenefit," which comes from British contract law, *Williams v. Roffey Bros. & Nicholls (Contractors) Ltd.* [1990] 1 All E.R. 512.

14. For citations for the quotations in this paragraph, as well as some longer quotations, all from the "Experience Project" site, see Elizabeth F. Emens, "Admin," *Georgetown Law Journal* 103, no. 6 (2015): 1454–55 and notes 191–94.
15. See, for example, *English Oxford Living Dictionaries*, "bureaucracy (*n.*)," https://en.oxforddictionaries.com/thesaurus/bureaucracy (listing *paperwork* as a synonym for *bureaucracy*).
16. See "Summer Camp Information," Arlington Parks and Recreation, https://parks.arlingtonva.us/programs/summer-camps.

## 2. The Costs of Admin, or Where's Your Head?

1. Pamela Paul, "Sleep Medication: Mother's New Little Helper," *New York Times*, November 4, 2011, https://www.nytimes.com/2011/11/06/fashion/mothers-and-sleep-medication.html; Joan Williams, "Erasmus B. Dragon: Inequality Is a Joke to the *New York Times*," *Huffington Post*, updated January 9, 2012, https://www.huffingtonpost.com/joan-williams/erasmus-b-dragon-inequali_b_1080744.html (responding to Paul's article and critiquing its lack of attention to women's primary role in household management).
2. See, for example, Lyn Craig and Killian Mullan, "Parental Leisure Time: A Gender Comparison in Five Countries," *Social Politics* 20, no. 3 (September 2013): 343–44, figure 3 and table 2; Liana C. Sayer et al., "How Long Is the Second (Plus First) Shift? Gender Differences in Paid, Unpaid, and Total Work Time in Australia and the United States," *Journal of Comparative Family Studies* 40, no. 4 (June 2009): 541. But see Melissa A. Milkie, Sara B. Raley, and Suzanne M. Bianchi, "Taking on the Second Shift: Time Allocations and Time Pressures of U.S. Parents with Preschoolers," *Social Forces* 88, no. 2 (December 2009): 500.
3. See David A. Rosenbaum, Lanyun Gong, and Cory Adam Potts, "Pre-Crastination: Hastening Subgoal Completion at the Expense of Extra Physical Effort," *Psychological Science* 25, no. 7 (May 2014).
4. On the distinction between picking and choosing, see Edna Ullmann-Margalit and Sidney Morgenbesser, "Picking and Choosing," *Social Research* 44, no. 4 (Winter 1977): 761–62; Cass R. Sunstein, "Choosing Not to Choose," *Duke Law Journal* 64 (2014): 12, note 31.
5. This insight exemplifies the "social model" of disability, which sees disability as the interaction between an impairment and the surrounding so-

cial environment, rather than as an individual medical problem. See, for example, Samuel R. Bagenstos, *Law and the Contradictions of the Disability Rights Movement* (New Haven, CT: Yale University Press, 2009), 18–20.

6. See Chapter 13 for more discussion of the legal dimension of lost personal time.
7. D. Lyman, Jr., trans., *The Moral Sayings of Publius Syrus, a Roman Slave: From the Latin* (New York: A. J. Graham, 1862), 13.
8. Erving Goffman, *Interaction Ritual: Essays in Face-to-Face Behavior* (New Brunswick, NJ: AldineTransaction, 2005), 133.
9. See Mullainathan and Shafir, *Scarcity*, 49–51.
10. Ibid., 13, 27–29, 36–37, 197.
11. See Bluma Zeigarnik, "On Finished and Unfinished Tasks," in *A Source Book of Gestalt Psychology*, ed. Willis D. Ellis (London: Kegan Paul, Trench, Trubner, 1938): 300–314; Timo Mäntylä and Teresa Sgaramella, "Interrupting Intentions: Zeigarnik-Like Effects in Prospective Memory," *Psychological Research* 60, no. 3 (September 1997): 197.
12. I thank Sendhil Mullainathan for this analogy.
13. See, for example, Mihaly Csikszentmihalyi, *Flow: The Psychology of Optimal Experience* (New York: Harper and Row, 1990).
14. See Shira Offer and Barbara Schneider, "Revisiting the Gender Gap in Time-Use Patterns: Multitasking and Well-Being Among Mothers and Fathers in Dual-Earner Families," *American Sociological Review* 76, no. 6 (December 2011): 828.
15. See, for example, L. Melissa Skaugset et al., "Can You Multitask? Evidence and Limitations of Task Switching and Multitasking in Emergency Medicine," *Annals of Emergency Medicine* 68, no. 2 (August 2016): 189–95.
16. Goleman, *Focus*, 5.
17. Credit for the "no toothache" idea goes to Thich Nhat Hanh. See "Experiencing Our Body" in his collection *The Path of Emancipation: Talks from a 21-Day Mindfulness Retreat* (Berkeley: Parallax, 2000), 78.
18. Susan Parker, "Esther Duflo Explains Why She Believes Randomized Controlled Trials Are So Vital," Center for Effective Philanthropy, June 23, 2011, http://cep.org/esther-duflo-explains-why-she-believes-randomized-controlled-trials-are-so-vital.
19. Kaitlyn Greenidge, "The Dread of Taxes That Even Beyoncé Can't Fix,"

*New York Times*, April 8, 2017, https://www.nytimes.com/2017/04/08/opinion/sunday/the-dread-of-taxes-that-even-beyonce-cant-fix.html.

20. See, for example, Alexes Harris, *A Pound of Flesh: Monetary Sanctions as Punishment for the Poor* (New York: Russell Sage Foundation, 2016); Wayne A. Logan and Ronald F. Wright, "Mercenary Criminal Justice," *University of Illinois Law Review* 2014, no. 4 (2014); Issa Kohler-Hausmann, *Misdemeanorland: Criminal Courts and Social Control in an Age of Broken Windows Policing* (Princeton, NJ: Princeton University Press, 2018); United States Department of Justice Civil Rights Division, Investigation of the Ferguson Police Department, March 4, 2015, https://www.justice.gov/sites/default/files/opa/press-releases/attachments/2015/03/04/ferguson_police_department_report.pdf.

21. On this case, see, for example, German Lopez, "The Tyranny of a Traffic Ticket: How Small Crimes Turn Fatal for Poor, Minority Americans," *Vox*, August 10, 2016, http://www.vox.com/2016/8/5/12364580/police-over criminalization-net-widening; Isaac Bailey, "In Castile Shooting, a 4-Year-Old Gives Her Mother 'the Talk,'" CNN, June 22, 2017, https://www.cnn.com/2017/06/22/opinions/castile-shooting-4-year-old-response-bailey/index.html.

## 3. Admin Personalities, or Who Are You?

1. "Pretty Obvious Which Sibling Going to Have to Deal with All the Nursing Home Stuff," *Onion*, December 4, 2013, http://www.theonion.com/article/pretty-obvious-which-sibling-going-to-have-to-deal-34744.

2. "401K Enrollment Form Sits at Bottom of Desk Drawer for 22 Years," *Onion*, December 12, 2001, http://www.theonion.com/graphic/401k-enrollment-form-sits-at-bottom-of-desk-drawer-8792.

3. "Man Waiting in H&R Block Lobby Nervously Eyeing How Much More Paperwork Everyone Else Brought," *Onion*, April 8, 2015, http://www.theonion.com/article/man-waiting-in-hr-block-lobby-nervously-eyeing-how-38393.

4. "Woman Going to Take Quick Break After Filling Out Name, Address on Tax Forms," *Onion*, April 3, 2014, https://www.theonion.com/woman-going-to-take-quick-break-after-filling-out-name-1819576310.

5. With thanks to Tina Fey via Liz Lemon, "Cougars," *30 Rock*, NBC, first broadcast November 29, 2007.

6. For Vera, this book has been meaningful in understanding Saul around admin (and other areas): Gary Chapman, *The Five Love Languages: The Secret to Love That Lasts* (Chicago: Northfield, 2015).
7. Anita Wyzanski Robboy, *Aftermarriage: The Myth of Divorce* (Indianapolis: Alpha Books, 2002).

## 4. Who Does Admin?, or Is Admin for Girls?

1. Mary Anne Case, "How High the Apple Pie? A Few Troubling Questions About Where, Why, and How the Burden of Care for Children Should Be Shifted," *Chicago-Kent Law Review* 76, no. 3 (2001), 1764, note 30 (quoting "Three Texans Who Find Themselves in Washington After the George W. Bush Victory in the Presidential Election," interview by Renee Montagne, NPR, January 19, 2001).
2. Lisa Belkin, "When Mom and Dad Share It All," *New York Times Magazine*, June 15, 2008.
3. United States Bureau of Labor Statistics, "American Time Use Survey—2016 Results," June 27, 2017, table 1, http://www.bls.gov/news.release/pdf/atus.pdf.
4. Ibid., 4; Anne E. Winkler and Thomas R. Ireland, "Time Spent in Household Management: Evidence and Implications," *Journal of Family Economic Issues* 30, no. 3 (September 2009): 297, 301–2.
5. Judith Treas and Tsui-o Tai, "How Couples Manage the Household: Work and Power in Cross-National Perspective," *Journal of Family Issues* 33, no. 8 (August 2012): 1089. For a discussion of definitions of household management and the subsets of life admin it includes, see Emens, "Admin," 1433, note 83, and 1434.
6. Anne C. Weisberg and Carol A. Buckler, *Everything a Working Mother Needs to Know* (New York: Doubleday, 1994), 133.
7. Treas and Tai, "How Couples Manage," 1089.
8. For example, see Shira Offer, "The Costs of Thinking About Work and Family: Mental Labor, Work-Family Spillover, and Gender Inequality Among Parents in Dual-Earner Families," *Sociological Forum* 29, no. 4 (December 2014): 924, 931 (finding no significant difference in how much "family-specific mental labor" mothers and fathers did but finding that the amount of family-specific mental labor was inversely correlated with

well-being for mothers but not fathers, and defining family-specific mental labor as "thoughts about family, children, and spouse").

9. Pew Research Center, "Raising Kids and Running a Household: How Working Parents Share the Load," November 2015, 3, http://assets.pewresearch.org/wp-content/uploads/sites/3/2015/11/2015-11-04_working-parents_FINAL.pdf. A skew toward women—though not a majority—also emerges for "taking care of the children when they're sick," with 47% saying "mother does more" and "share equally." Caring for sick kids is likely a combination of direct care and admin because of the juggling involved in taking time off work to care for sick kids or the associated responsibility of finding alternate care if you're not available to stay home.

10. For an interesting related finding of reporting differences, see Michelle L. Frisco and Kristi Williams, "Perceived Housework Equity, Marital Happiness, and Divorce in Dual-Earner Households," *Journal of Family Issues* 24, no. 1 (January 2003): 64, 68.

11. Pew Research Center, "Raising Kids," 11.

12. Claire Cain Miller, "Stressed, Tired, Rushed: A Portrait of the Modern Family," *New York Times*, November 4, 2015, http://www.nytimes.com/2015/11/05/upshot/stressed-tired-rushed-a-portrait-of-the-modern-family.html.

13. See, for example, Jo A. Meier, Mary McNaughton-Cassill, and Molly Lynch, "The Management of Household and Childcare Tasks and Relationship Satisfaction in Dual-Earner Families," *Marriage and Family Review* 40, nos. 2/3 (2006): 70–71, table 1, 72–73, table 2; H. Wesley Perkins and Debra DeMeis, "Gender and Family Effects on the 'Second-Shift' Domestic Activity of College-Educated Young Adults," *Gender and Society* 10 (1996): 86 (finding that women working full-time were significantly more likely to engage in meal planning when married with kids than when married with no kids, whereas men were less likely to do so after kids); "Who Does What in the Modern Family Home," Mumsnet.com, https://www.mumsnet.com/surveys/chores-survey-results; Hochschild and Machung, *Second Shift*, 38, 79, 258; Marjorie L. DeVault, *Feeding the Family: The Social Organization of Caring as Gendered Work* (Chicago: University of Chicago Press, 1991). Note also that a woman need not be partnered with a man to do more kidmin than the men around her because, for instance, children are more likely to live only with a mother than with a father. See Emens, "Admin," 1437.

14. On social admin within families, or kinwork, see Micaela di Leonardo, "The Female World of Cards and Holidays: Women, Families, and the Work of Kinship," *Signs* 12, no. 3 (Spring 1987): 442–43.
15. See Meier, McNaughton-Cassill, and Lynch, "The Management of Household and Childcare Tasks," 70, table 1, 72, table 2; Helen J. Mederer, "Division of Labor in Two-Earner Homes: Task Accomplishment Versus Household Management as Critical Variables in Perceptions About Family Work," *Journal of Marriage and Family* 55, no. 1 (February 1993): 139; Perkins and DeMeis, "Gender and Family Effects," 86. On the care work of "feeding the family," see DeVault, *Feeding the Family*; Sandra Colavecchia, "Moneywork: Caregiving and the Management of Family Finances," in *Family Patterns: Gender Relations*, ed. Bonnie Fox, 3rd ed. (Don Mills, Ontario: Oxford University Press, 2009), 422.
16. See Meier, McNaughton-Cassill, and Lynch, "The Management of Household and Childcare Tasks," 70, table 1, 72, table 2; Mederer, "Division of Labor," 139.
17. See, for example, Elizabeth Warren and Amelia Warren Tyagi, *The Two-Income Trap: Why Middle-Class Parents Are Going Broke* (New York: Basic Books, 2003), 11; Deborah Thorne, "Extreme Financial Strain: Emergent Chores, Gender Inequality and Emotional Distress," *Journal of Family and Economic Issues* 31, no. 2 (June 2010): 187, 194–95; Kenneth Matos, "Modern Families: Same- and Different-Sex Couples Negotiating at Home," Families and Work Institute, 2015, 13.
18. See Meier, McNaughton-Cassill, and Lynch, "The Management of Household and Childcare Tasks," 70, table 1, 72, table 2.
19. Ibid.
20. See, for example, "Who Does What in the Modern Family Home"; Meier, McNaughton-Cassill, and Lynch, "The Management of Household and Childcare Tasks," 72–73, table 2 (not including data on who finds or manages childcare but reporting other measures of childcare management as opposed to direct childcare labor in their study that suggest that the gender gap is even greater on the management front than the task front for stereotypically female tasks).
21. See, for example, Liza Mundy, "The Gay Guide to Wedded Bliss," *Atlantic*, June 2013, http://www.theatlantic.com/magazine/archive/2013/06/the-gay-guide-to-wedded-bliss/309317; Charlotte J. Patterson and Rachel H. Farr, "Coparenting Among Lesbian and Gay Couples," in *Copar-*

*enting*, ed. James P. McHale and Kristin M. Lindahl (Washington, DC: American Psychological Association, 2011), 131, 134.

22. See, for example, Christopher Carrington, *No Place Like Home: Relationships and Family Life Among Lesbians and Gay Men* (Chicago: University of Chicago Press, 1999), 158–59.
23. See, for example, Suzanne Taylor Sutphin, "Social Exchange Theory and the Division of Household Labor in Same-Sex Couples," *Marriage and Family Review* 46, no. 3 (June 2010): 202. Another confirms that the lower-earning spouse in a same-sex couple is unlikely to handle investments, but the job is almost equally likely to be shared (37%) as to fall to the higher earner (40%). Matos, "Modern Families," 14.
24. See Scott Coltrane and Kristy Y. Shih, "Gender and the Division of Labor," in *Handbook of Gender Research in Psychology*, ed. Joan C. Chrisler and Donald R. McCreary (New York: Springer Science, 2010), 411; see also Mignon R. Moore, "Gendered Power Relations Among Women: A Study of Household Decision Making in Black, Lesbian Stepfamilies," *American Sociological Review* 73, no. 2 (April 2008): 345–46. The studies often talk about "the biological mother," which is an ambiguous phrase, since you can contribute to a child's biology in multiple ways, but I take them to mean the partner who gestates the baby.
25. See, for example, Gayle Rubin, "The Traffic in Women," *Toward an Anthropology of Women*, ed. Rayna Reiter (New York: Monthly View Press, 1975): 157–210.
26. Carrington, *No Place Like Home*, 14.
27. Rose Hackman, "'Women Are Just Better at This Stuff': Is Emotional Labor Feminism's Next Frontier?" *Guardian*, November 8, 2015, https://www.theguardian.com/world/2015/nov/08/women-gender-roles-sexism-emotional-labor-feminism. The article defines "emotional labor" to include things like remembering birthdays and planning mealtimes with family members' preferences and schedules in mind and thus as overlapping with life admin; it also recognizes Arlie Hochschild's key role in defining the concept. See Arlie Hochschild, *The Managed Heart: Commercialization of Human Feeling*, 3rd ed. (Berkeley: University of California Press, 2012), 23, 38–49.
28. Andre J. Szameitat et al., "'Women Are Better Than Men'—Public Beliefs on Gender Differences and Other Aspects in Multitasking," *PLoS One* 10, no. 10 (October 2015): 10.

29. David Z. Hambrick et al., "Predictors of Multitasking Performance in a Synthetic Work Paradigm," *Applied Cognitive Psychology* 24, no. 8 (September 2010): 1164; Timo Mäntylä, "Gender Differences in Multitasking Reflect Spatial Ability," *Psychological Science* 24, no. 4 (April 2013): 518–19.
30. See, for example, Szameitat et al., "'Women Are Better,'" 8; Suzanne M. Bianchi and Vanessa R. Wight, "The Long Reach of the Job: Employment and Time for Family Life," in *Workplace Flexibility: Realigning 20th-Century Jobs for a 21st-Century Workforce*, ed. K. Christensen and B. Schneider (Ithaca, NY: Cornell University Press, 2010), 39; Offer and Schneider, "Revisiting the Gender Gap," 813, 821, 828.
31. Brent A. McBride et al., "Paternal Identity, Maternal Gatekeeping, and Father Involvement," *Family Relations* 54, no. 3 (July 2005): 362 (internal quotation marks omitted). See Scott Coltrane, *Family Man: Fatherhood, Housework, and Gender Equity* (Oxford Univ. Press: 1996), 74–76.
32. Belkin, "When Mom and Dad Share It All." See also Treas and Tai, "How Couples Manage," 1097, 1109–11 (finding that making more money correlates with women, but not men, having more control over household decision making, across a multinational sample, while also finding that most couples in all countries report sharing the decisions, in a study that asks only about "who makes decisions" and who has the "final say"). See also Moore, "Gendered Power Relations," 346 (reporting on a "modest boasting of their superior cleaning and organizational skills" by "biological mothers" in African-American lesbian stepfamilies).
33. Our topic is life admin rather than work admin, but it's worth mentioning here the fascinating literature on gender and admin at work, including the studies finding that women do more (unrewarded) admin than men in similar jobs and that when women assume leadership roles at work, the jobs lose status, as if the jobs themselves are considered more secretarial simply because women are doing them. For some citations, see my article "Admin," 1438.
34. For an argument for family "fairness" as preferable to a vision of family "equality" per Susan Okin, see Edna Ullmann-Margalit, "Family Fairness," *Social Research* 73, no. 2 (Summer 2006): 575.
35. Marjorie DeVault has nicely observed, as to key elements of household management, that its "invisibility" can make it "difficult to share . . . because one never knows whether another is thinking about what is needed,"

and because, for some tasks, "maintaining their invisibility is part of doing the work well, [so] people are often unable, or reluctant, to talk explicitly about them." DeVault, *Feeding the Family*, 141.



## 5. Admin Is Sticky, or If Everybody's Doing It, How Come Some People Are Doing More of It?

1. See, for example, Sue Shellenbarger, " 'What's the Netflix Password Again, Mom?,' " *Wall Street Journal*, March 13, 2013, https://www.wsj.com/articles/SB10001424127887324128504578348613932711322 (reporting that "more than 2 in 5 parents of 18- to 35-year old children still pay for their kids' cellphone service" and 29% still do so for kids who don't live at home or get help with rent). Though the financial costs to parents are typically the focus of these discussions, this dynamic involves admin costs as well.
2. See, for example, Eric J. Johnson, Mary Steffel, and Daniel G. Goldstein, "Making Better Decisions: From Measuring to Constructing Preferences," *Health Psychology* 24, no. 4 (2005): S18–S19; Brigitte C. Madrian and Dennis F. Shea, "The Power of Suggestion: Inertia in 401(k) Participation and Savings Behavior," *Quarterly Journal of Economics* 116, no. 4 (November 2001): 1176.
3. See, for example, Johnson, Steffel, and Goldstein, "Making Better Decisions," S18.
4. Margaret Joan Anstee, *Never Learn to Type: A Woman at the United Nations* (New York: Wiley and Sons, 2004), xii.
5. For an example of successfully redistributing the snail-mail job, see Dufu, *Drop the Ball*, 113–20.
6. Nina Totenberg, "Justice Ruth Bader Ginsburg Reflects on the #MeToo Movement: 'It's About Time,' " NPR, January 22, 2018, https://www.npr.org/2018/01/22/579595727/justice-ginsburg-shares-her-own-metoo-story-and-says-it-s-about-time.

## 6. Admin That Can Wreck You

1. Roz Chast, *Can't We Talk About Something More Pleasant?* (New York: Bloomsbury, 2014), 86.

2. Marge Piercy, "To Be of Use," in *To Be of Use* (Garden City, NY: Doubleday, 1973), 49.
3. Dorothy A. Miller, "The 'Sandwich' Generation: Adult Children of the Aging," *Social Work* 26, no. 5 (September 1981).
4. On care-work in general, see, for example, Kerstin Aumann et al., "Working Family Caregivers of the Elderly: Everyday Realities and Wishes for Change," Families and Work Institute, 2010, 6; Ann Bookman and Delia Kimbrel, "Families and Elder Care in the Twenty-First Century," *Future of Children* 21, no. 2 (2011): 125. Studies generally focus more on direct care, but some work suggests that the admin-doing is also shaped by race or ethnicity, possibly in more complicated ways. See, for example, National Alliance for Caregiving and AARP, "Caregivers of Older Adults: A Focused Look at Those Caring for Someone Age 50+" (June 2015): 1, 8, 23, 24, https://www.aarp.org/content/dam/aarp/ppi/2015/caregivers-of-older-adults-focused-look.pdf (reporting on greater elder caregiving generally in Asian-American, Hispanic, and African-American communities and also specifically reporting that "Asian-American caregivers are more likely . . . than all other caregivers of other race/ethnicities to communicate with health care professionals on behalf of their care recipient").
5. Hochschild, *The Managed Heart*, 23, 38–49.
6. See Rachel Adams, *Raising Henry: A Memoir of Motherhood, Disability, and Discovery* (New Haven, CT: Yale University Press, 2013), 82.
7. A number of states and localities prohibit landlords from refusing tenants who rely on federally subsidized (Section 8) housing vouchers, although federal law permits this discrimination. See, for example, Poverty and Race Research Action Council, *Expanding Choice: Practical Strategies for Building a Successful Housing Mobility Program* Appendix B (August 2017), http://www.prrac.org/pdf/AppendixB.pdf.
8. Shonda Rhimes, *Year of Yes: How to Dance It Out, Stand in the Sun and Be Your Own Person* (New York: Simon and Schuster, 2015), 232.
9. Gretchen Rubin, "My New Habit for Tackling Nagging Tasks: Power Hour" (February 21, 2014), https://gretchenrubin.com/2014/02/my-new-habit-for-tackling-nagging-tasks-power-hour (describing this "Secret of Adulthood").
10. Kimberly Norwood, "Why I Fear for My Sons," CNN, August 25, 2014, http://www.cnn.com/2014/08/25/opinion/norwood-ferguson-sons-brown-police/index.html. These kinds of experiences lead some black parents to consider whether to send their children to programs with titles

like "Race & the Law" and "Surviving the Stop," which critics say teach children to ignore their individual rights and proponents say teach necessary survival strategies. See, for example, Janell Ross, "Black Parents Take Their Kids to School on How to Deal with Police," *Washington Post*, January 3, 2017, https://www.washingtonpost.com/national/black-parents-take-their-kids-to-school-on-how-to-deal-with-police/2017/01/03/86129c1c-c6be-11e6-bf4b-2c064d32a4bf_story.html.

## 7. Admin That Can Fix You

1. Stephen R. Covey, *The 7 Habits of Highly Effective People: Restoring the Character Ethic* (New York: Simon and Schuster, 2004), 165.
2. This chart is adapted from ibid., 160.
3. Ibid., 165.
4. Ruth Davis Konigsberg, "Women and Time: What Makes Us Tick," *Real Simple* (April 2012), 64 (reporting on a poll finding that "50% of women who schedule their free time regularly are satisfied with their lives but 41% of the women who postpone their free time until they finish other tasks are satisfied with their lives").
5. Richard H. Thaler and Cass R. Sunstein, *Nudge: Improving Decisions About Health, Wealth, and Happiness* (New Haven, CT: Yale University Press, 2008).
6. Khe Hy, "Cure Your Bad Habits with 'Sludge,' the Productivity Tip You've Never Hea[r]d Of," Quartz at Work, January 22, 2018, https://work.qz.com/1181942/be-a-better-worker-with-sludge-the-behavioral-science-productivity-trick. Arguably much of what's called sludge is covered by the idea of nudge (since you can nudge in any direction), but *sludge* is a great word for when the gentle imposition gets in the way of something you're tempted to do.
7. Ibid. The password example that follows is also from this source.
8. Hengchen Dai, Katherine L. Milkman, and Jason Riis, "The Fresh Start Effect: Temporal Landmarks Motivate Aspirational Behavior," *Management Science* (2014), https://faculty.wharton.upenn.edu/wp-content/uploads/2014/06/Dai_Fresh_Start_2014_Mgmt_Sci.pdf.
9. This also included the ten- to twenty-minute-per-week exercise regime from this book: Jonathan Bailor, *The Calorie Myth: How to Eat More, Exercise Less, Lose Weight, and Live Better* (New York: HarperCollins, 2014).

10. This question brings together Vera's insight with a question the psychologist and meditation teacher Tara Brach urges people who are feeling angry or suffering from painful repetitive thoughts to ask themselves: *What would I have to feel or know if I weren't feeling angry right now?* For a similar formulation, see, for example, https://www.tarabrach.com/taking-exquisite-risk-undefended-heart-transcript.

## 8. Admin Judgments

1. I strongly disfavor using words, like *crazy*, that are pejorative terms for disability. I have not thought of Carlin's reference to idiots and maniacs in relation to mental disability, though of course these terms do have that deeply problematic history; I have not made the association, probably because, unlike *crazy* and *nuts*, the terms *idiots* and *maniacs* are not currently in use in my communities as terms for actual mental disability. This is not a satisfying explanation, however. For a longer discussion of these important questions of language and disability discrimination, see, for example, Joseph P. Shapiro, *No Pity: People with Disabilities Forging a New Civil Rights Movement* (New York: Times Books, 1993), and Rachel Cohen-Rottenberg, "Doing Social Justice: 10 Reasons to Give up Ableist Language," *Huffington Post*, June 10, 2014, https://www.huffingtonpost.com/rachel-cohenrottenberg/doing-social-justice-thou_b_5476271.html.
2. Woody Allen has said that his actual line was "80 percent of success is showing up"; William Safire, "On Language; The Elision Fields," *New York Times Magazine*, August 13, 1989, https://www.nytimes.com/1989/08/13/magazine/on-language-the-elision-fields.html.
3. Dan Harris, *10% Happier: How I Tamed the Voice in My Head, Reduced Stress Without Losing My Edge, and Found Self-Help That Actually Works* (New York: Dey Street Books, 2014), xiv.
4. To be clear, I do not think that the meditation or yoga traditions that I know treat obligations to other people as insignificant or unreal. On the contrary, ethics and treatment of others are central concerns. What I am describing is the music—the tone or implication—of the words I have heard from some minority of teachers I have encountered. (Perhaps I have misinterpreted them, in which case I hope that writing these words might help to bridge that gap.)
5. Bill Thornton et al., "The Mere Presence of a Cell Phone May Be Distracting," *Social Psychology* 45, no. 6 (November 2014): 483–85.

6. Even at places that say they have a phone number for emergency messages to reach you, on arrival I've been told that the emergency line gets checked only once or twice some days. When that hasn't been enough for my sense of obligation, I've been told I just need to keep my phone on—contrary to the rules of the retreat—without regard to the possible annoyance or judgment of other participants.
7. People's views of what is (or isn't) necessary may also change over time or in response to relational pressure. Hochschild found that "men whose wives pressed them to do more often resisted by reducing their ideas about needs." Hochschild and Machung, *Second Shift*, 252.
8. For discussion, and more renditions, see https://quoteinvestigator.com/2010/06/29/be-kind. My favorite version of the line, which I've quoted, appears in Elizabeth Coleman's poem "Imagine Peace," in Elizabeth J. Coleman, *The Fifth Generation* (New York: Spuyten Duyvil, 2016), 5

## 9. Admin Pleasures

1. Leo Bersani, "Is the Rectum a Grave?," *AIDS: Cultural Analysis/Cultural Activism* 43 (Winter 1987): 197.
2. Daniel Kahneman et al., "A Survey Method for Characterizing Daily Life Experience: The Day Reconstruction Method," *Science* 306 (2004): 1777.
3. On asexuality, see, for example, Elizabeth F. Emens, "Compulsory Sexuality," *Stanford Law Review* 66, no. 2 (February 2014).
4. On heuristics and misfiring beyond the realm of simple logical errors, see, for example, Cass R. Sunstein, "Moral Heuristics," *Behavioral and Brain Sciences* 28, no. 4 (August 2005).
5. A vast and fascinating literature explores terms like *pleasure, satisfaction, desire*, and more, the distinctions among which are overlooked in this discussion.
6. Julia Cameron, *The Artist's Way: A Spiritual Path to Higher Creativity* (New York: Tarchur Putnam, 2002).
7. See Matthew B. Crawford, *The Case for Working with Your Hands, or, Why Office Work Is Bad for Us and Fixing Things Feels Good* (London: Penguin Viking, 2001).
8. Linda Babcock and Sara Laschever, *Women Don't Ask: Negotiation and the Gender Divide* (Princeton, NJ: Princeton University Press, 2003); Linda Babcock and Sara Laschever, *Ask for It: How Women Can Use the Power*

*of Negotiation to Get What They Really Want* (New York: Bantam, 2009). The prospect of being discriminated against, supported by data in some domains, may also affect one's preferences (or aversions). See Case, "Developing a Taste."

## 10. Admin to Win Friends and Influence People

1. See, for example, Barbara Frederickson, *Positivity: Top-Notch Research Reveals the Upward Spiral That Will Change Your Life* (New York: Three Rivers, 2009), 41–42, 92–93, 186–87.
2. Jackie Calmes, "Activists Help Pay for Patients' Travel to Shrinking Number of Abortion Clinics," *New York Times*, November 27, 2014, https://www.nytimes.com/2014/11/28/us/advocates-help-pay-for-travel-to-a-shrinking-number-of-abortion-clinics.html. For a discussion of the Supreme Court's decision striking down Texas's House Bill 2, which caused the clinic closures that inspired the organization Sheible cofounded (Fund Texas Choice), and of Texas's subsequent law in the context of a broader analysis of US abortion debates, see Carol Sanger, *About Abortion: Terminating Pregnancy in Twenty-First-Century America* (Cambridge, MA: Belknap Press, 2017), 34–36, 235.
3. See, for example, Kathleen Kingsbury, "Longer Wait Times, Higher Costs for U.S. Adoptions," Reuters, January 15, 2018, https://www.reuters.com/article/us-adoption-domestic-waits-idUSBRE90E15Y20130115; Steve Deace, "The Red Tape Around Adoption," *USA Today*, May 6, 2013, http://www.usatoday.com/story/opinion/2013/05/03/adoption-stuck-bureaucracy-column/2124199.
4. See, for example, Jessan Hutchinson-Quillian, David Reiley, and Anya Samek, "Hassle Costs and Workplace Charitable Giving: Field Experiments with Google Employees," https://papers.ssrn.com/sol3/papers.cfm?abstract_id=3204588# (posted July 18, 2018), and studies cited therein.
5. See, for example, Hannah Riley Bowles and Linda Babcock, "How Can Women Escape the Compensation Negotiation Dilemma? Relational Accounts Are One Answer," *Psychology of Women Quarterly* 37, no. 1 (2012).
6. Sijun Wang, Sharon E. Beatty, and Jeanny Liu, "Employees' Decision Making in the Face of Customers' Fuzzy Return Requests," *Journal of Marketing* 76, no. 6 (November 2012): 80.
7. Whether, in general, a partner's doing more admin will affect how much sex the partners have is an open question, empirically. Quantitative data

on the relationship between sex and housework are mixed and in need of further study (especially for same-sex couples) and, as far as I can tell, unavailable with regard to admin-doing in particular. Those interested in the subject might start with this recent study: Matthew D. Johnson, Nancy L. Galambos, and Jared R. Anderson, "Skip the Dishes? Not So Fast! Sex and Housework Revisited," *Journal of Family Psychology* 30, no. 2 (2016): 208; for a recent study suggesting that the personality trait of conscientiousness may be linked to sexual satisfaction, however, see Julia Velten, Julia Brailovskaia, and Jürgen Margraf, "Exploring the Impact of Personal and Partner Traits on Sexuality: Sexual Excitation, Sexual Inhibition, and Big Five Predict Sexual Function in Couples," *Journal of Sex Research* (2018).

## 11. Relationship Tips

1. Murray R. Barrick and Michael K. Mount, "The Big Five Personality Dimensions and Job Performance: A Meta-Analysis," *Personnel Psychology* 44, no. 1 (March 1991): 4 (noting this as one understanding of what conscientiousness entails).
2. Brittany C. Solomon and Joshua J. Jackson, "The Long Reach of One's Spouse: Spouses' Personality Influences Occupational Success," *Psychological Science* 25, no. 12 (December 2014): 2194.
3. Ibid., 2195; Portia S. Dyrenforth et al., "Predicting Relationship and Life Satisfaction from Personality in Nationally Representative Samples from Three Countries: The Relative Importance of Actor, Partner, and Similarity Effects," *Journal of Personality and Social Psychology* 99, no. 4 (2010): 700.
4. Jason K. White, Susan S. Hendrick, and Clyde Hendrick, "Big Five Personality Variables and Relationship Constructs," *Personality and Individual Differences* 37, no. 7 (November 2004): 1523, 1525, 1527.
5. Julia Velten, Julia Brailovskaia, and Jürgen Margraf, "Exploring the Impact of Personal and Partner Traits on Sexuality: Sexual Excitation, Sexual Inhibition, and Big Five Predict Sexual Function in Couples," *Journal of Sex Research* (2018): 7 (reporting on results of a study with 98% different-sex couples in "steady" relationships).
6. Nora Ephron, *I Feel Bad About My Neck* (New York: Vintage Books, 2008), 123.
7. For more discussion of polyamory, see Elizabeth F. Emens, "Monogamy's

Law: Compulsory Monogamy and Polyamorous Existence," *NYU Review of Law and Social Change* 29, no. 2 (2004).

8. For a discussion of how much admin accompanies name changes at marriage, see Emens, "Changing Name Changing," 809–13. On the more onerous admin faced by trans people who want to change name and gender, see Dean Spade, *Normal Life: Administrative Politics, Critical Trans Politics, and the Limits of Law* (Durham, NC: Duke University Press, 2015), 79.
9. Matos, "Modern Families," 18.
10. Erich Fromm, *The Art of Listening* (New York: Continuum International, 2009), 192–93.
11. Alan J. Hawkins et al., "An Evaluation of a Program to Help Dual-Earner Couples Share the Second Shift," *Family Relations* 43 (April 1994): 217–19.
12. Ibid., 216–17, 219.
13. Allen, *Getting Things Done*, 13–14.
14. On reciprocity, see, for example, Robert B. Cialdini, *Influence: Science and Practice* (Boston: Allyn and Bacon, 2001), 20–50.
15. Credit for the excellent term *economy of gratitude* goes to Hochschild and Machung, *Second Shift*, 18.
16. The passage draws inspiration from Rilke's famous remarks on marriage. See Rainer Maria Rilke, *Rilke on Love and Other Difficulties*, trans. John J. L. Mood (New York: W. W. Norton, 2004), 34.

## 12. Individual Strategies

1. Ashley Craig et al., "Waiting to Give," IZA discussion paper no. 8491, Institute for the Study of Labor, Bonn, Germany (September 2014): 11–12, http://ssrn.com/abstract=2505353.
2. Barry Schwartz, *The Paradox of Choice: Why More Is Less* (New York: Ecco, 2004); Barry Schwartz et al., "Maximizing Versus Satisficing: Happiness Is a Matter of Choice," *Journal of Personality and Social Psychology* 83, no. 5 (November 2002).
3. See Marie Kondo, *The Life-Changing Magic of Tidying Up: The Japanese Art of Decluttering and Organizing* (New York: Ten Speed Press, 2014).
4. For a discussion of how an Amazon customer can lose his proverbial rightness, see Tuan Do, "Return Too Many Items on Amazon? Will Amazon Ban Your Account?," Techwalls, updated October 25, 2017, https://www.techwalls.com/amazon-ban-return-too-many-items.

5. Helen Fielding, *Bridget Jones's Diary* (London: Picador, 1996), 89.
6. I've heard this advice attributed to Harriet Lerner, but the closest I've found in Lerner is the equally wise counsel that "change occurs slowly in close relationships. If you make even a small change, you will be tested many times to see if you 'really mean it.'" Harriet Lerner, *The Dance of Anger: A Woman's Guide to Changing the Patterns of Intimate Relationships* (New York: Perennial Library, 1986), 201.
7. Adapted from this retelling: "Use Metaphors and Analogies to Fire Up Imagination and Illumination," Peter Guber, https://www.peterguber.com/jack-warner. For another famous variation, see William Oncken Jr. and Donald L. Wass, "Management Time: Who's Got the Monkey?," *Harvard Business Review* (November/December 1974).
8. See Patricia J. Williams, *The Alchemy of Race and Rights* (Cambridge, MA: Harvard University Press, 1991), 146–48.
9. Annette Lareau, *Unequal Childhoods: Class, Race, and Family Life* (Berkeley: University of California Press, 2003), 238.
10. Henri J. M. Nouwen, *Life of the Beloved: Spiritual Living in a Secular World* (New York: Crossroad, 2015), 106.

## 13. Collective Possibilities

1. See David Frisch, "It's About Time," *Tennessee Law Review* 79, no. 4 (Summer 2012): 758–60; Leonard E. Gross, "Time and Tide Wait for No Man: Should Lost Personal Time Be Compensable?," *Rutgers Law Journal* 33, no. 3 (Spring 2002): 684. There are exceptions, though. See, for example, Gross, "Time and Tide," 685.
2. *In re* Hannaford Bros. Co. Customer Data Sec. Breach Litig., 613 F. Supp. 2d 108, 134 (D. Me. 2009).
3. See Frisch, "It's About Time," 792; Gross, "Time and Tide," 685.
4. "About the Financial Ombudsman Service," Financial Ombudsman Service, http://www.financial-ombudsman.org.uk/about/index.html; Financial Services and Markets Act 2000, Part XVI § 229(3)(b) (UK); "Complaint," Financial Conduct Authority, at (2)(a), https://www.handbook.fca.org.uk/handbook/glossary/G197.html.
5. 18 U.S.C. § 3663(b)(6) (2012); see also Frisch, "It's About Time," 767. Recovery would be up to the federal government, since there is no private right of action under the act. And DOJ reports no data on successful recov-

ery of funds under the Act, despite the many victims reported. Bureau of Justice Statistics, "17.6 Million U.S. Residents Experienced Identity Theft in 2014," https://www.bjs.gov/content/pub/press/vit14pr.cfm.



6. For alternative approaches to calculating these damages, see Frisch, "It's About Time," 792–801; Gross, "Time and Tide," 700–701. For a fuller discussion, see Emens, "Admin," 1476–77.
7. This works on Amazon unless or until you return too many items. See Do, "Return Too Many Items on Amazon?"
8. On comparison friction, see Cass R. Sunstein, *Simpler: The Future of Government* (New York: Simon and Schuster, 2013), 81, 86–87.
9. See *Wegelin v. Reading Hospital and Medical Center*, 909 F. Supp. 2d 421 (E.D. Pa. 2012), which concluded that a hospital employee's taking time off to find a suitable daycare for her autistic daughter after the hospital moved her parking spot far enough away that she couldn't pick up her daughter on time at the current daycare constituted "caring for" a family member with "a serious health condition" under the FMLA. The FMLA covers employers who have at least fifty employees within a seventy-five-mile radius, and employees who have worked at least 1,250 hours in the previous twelve months. See U.S. Department of Labor, "Family and Medical Leave Act (FMLA)," https://www.dol.gov/general/topic/workhours/fmla.
10. Many employers do provide at least some pay for FMLA-type leave. See Department of Labor, "Family Medical Leave in 2012: Executive Summary," ii, https://www.dol.gov/asp/evaluation/fmla/FMLA-2012-Executive-Summary.pdf.
11. For a discussion of rating programs and an empirical critique of their implementation, see Daniel E. Ho, "Fudging the Nudge: Information Disclosure and Restaurant Grading," *Yale Law Journal* 122 (2012): 574
12. See, for example, Yannis Bakos, Florencia Marotta-Wurgler, and David R. Trossen, "Does Anyone Read the Fine Print? Consumer Attention to Standard-Form Contracts," *Journal of Legal Studies* 43, no. 1 (January 2014): 32.
13. See Omri Ben-Shahar, "The Myth of the 'Opportunity to Read' in Contract Law," *ERCL* 1 (2009): 15–19.
14. Aleecia M. McDonald and Lorrie Faith Cranor, "The Cost of Reading Privacy Policies," *I/S: A Journal of Law and Policy for the Information Society* 4, no. 3 (2008): 563, https://www.cylab.cmu.edu/_files/pdfs/news/CostofReading.PDF.

15. See Joseph M. Perillo, *Corbin on Contracts* 7 (rev. ed. 2002), § 29.8.
16. See, for example, Margaret Jane Radin, *Boilerplate: The Fine Print, Vanishing Rights and the Rule of Law* (Princeton, NJ: Princeton University Press, 2012), 13–15, 226–32; Ronald J. Gilson, Charles F. Sabel, and Robert E. Scott, "Text *and* Context: Contract Interpretation as Contract Design," *Cornell Law Review* 100, no. 1 (November 2014): 75. My argument applies only to consumer contracts; individually negotiated contracts are, of course, different.
17. See, for example, Perillo, *Corbin on Contracts*, §§ 29.10–29.12, *Williams v. Walker-Thomas Furniture Company*, 350 F.2d 445 (D.C. Cir. 1965).
18. For related arguments, see, for example, Ian Ayres and Alan Schwartz, "The No-Reading Problem in Consumer Contract Law," *Stanford Law Review* 66 (March 2014): 545 (proposing a distinctive method for determining which terms should be inside the box); Ben-Shahar, "The Myth," 25–26 (arguing for "labeling" of consumer contracts akin to nutrition labeling).
19. The idea for a warning heading for the box comes from Ayres and Schwartz, "The No-Reading Problem," 601.
20. See, for example, Meirav Furth-Matzkin, "On the Unexpected Use of Unenforceable Contract Terms: Evidence from the Residential Rental Market," *Journal of Legal Analysis* 9, no. 1 (Spring 2017): 24. For data on how often various terms appear in consumer contracts with internet retailers and how often the contracts themselves are unenforceable, see Ronald J. Mann and Travis Siebeneicher, "Just One Click: The Reality of Internet Retail Contracting," *Columbia Law Review* 108 (2008): 999–1000, 1011.
21. See, for example, Bailey Kuklin, "On the Knowing Inclusion of Unenforceable Contract and Lease Terms," *University of Cincinnati Law Review* 56, no. 3 (1988): 845–47.
22. A few model laws are paving the way, though even there we need stronger penalties and more enforcement. See, for example, Tex. Bus. & Com. Code Ann. § 17.46(b)(12); D.C. ST. § 28–3904(e)(1); Furth-Matzkin, "On the Unexpected Use," 43; Kuklin, "On the Knowing Inclusion," 846. For a contrary argument, see Omri Ben-Shahar, "Fixing Unfair Contracts," *Stanford Law Review* 63 (2011): 869.
23. For further ideas, see, for example, Ben-Shahar, "The Myth," 21–26; Oren Bar-Gill, *Seduction by Contract: Law, Economics, and Psychology in Consumer Contracts* (Oxford: Oxford University Press, 2012).
24. Jay M. Feinman, *Delay, Deny, Defend: Why Insurance Companies Don't*

*Pay Claims and What You Can Do About It* (New York: Penguin, 2010), 58–61.

25. Ibid., 26, 30.

26. See, for example, Peter K. Stris, "ERISA Remedies, Welfare Benefits, and Bad Faith: Losing Sight of the Cathedral," *Hofstra Labor and Employment Law Journal* 26, no. 2 (January 2009): 393–95.

27. For an example of some of this information being collected, but not disclosed, see Feinman, *Delay, Deny, Defend*, 216–17.

28. For some ideas on this front, see, for example, Katherine T. Vukadin, "Delayed and Denied: Toward an Effective ERISA Remedy for Improper Processing of Healthcare Claims," *Yale Journal of Health Policy, Law, and Ethics* 11, no. 2 (2011): 334–36, 360–70; Feinman, *Delay, Deny, Defend*, 205–16, 218–22.

29. Jon Kabat-Zinn, *Wherever You Go, There You Are* (New York: Hyperion, 1994).

30. Kristin N. Ray et al, "Opportunity Costs of Ambulatory Medical Care in the United States," *American Journal of Managed Care* 21, no. 8 (August 2015): 571.

31. For important work on vulnerability and social policy, see, for example, Martha Fineman, "The Vulnerable Subject: Anchoring Equality in the Human Condition," *Yale Journal of Law and Feminism*, no. 20(1), (2012): 1.

32. Jonathan Cohn, "I'm Insanely Jealous of Sweden's Work-Family Policies. You Should Be Too," *New Republic* (June 22, 2014), https://newrepublic.com/article/118294; "Education in Sweden," Swedish Institute (last updated January 10, 2018), https://sweden.se/society/education-in-sweden.

33. Joan Williams, *Unbending Gender: Why Family and Work Conflict and What to Do About It* (New York: Oxford University Press, 2000), 55; Pamela Druckerman, *Bringing Up Bébé: One American Mother Discovers the Wisdom of French Parenting* (New York: Penguin, 2012), 104–8, 114, 215.

34. Schulte, *Overwhelmed*, 107–8; Karen Jowers, "More Bases Offering Extended Childcare Hours," *Military Times*, October 6, 2017, https://www.militarytimes.com/pay-benefits/2017/10/06/more-bases-offering-extended-child-care-hours; Druckerman, *Bringing Up Bébé*, 105.

35. See, for example, Eyal, Romain, and Robertson, "Can Rationing Through Inconvenience Be Ethical?" The authors acknowledge many of the problems noted in the rest of the paragraph.

36. Allison Daminger et al., "Poverty Interrupted: Applying Behavior Science

to the Context of Chronic Scarcity," *ideas42*, 2015, 27–28, http://www.ideas42.org/wp-content/uploads/2015/05/I42_PovertyWhitePaper_Digital_FINAL-1.pdf.



37. Ibid., 34.

38. See Mario Luis Small, *Unanticipated Gains: Origins of Network Inequality in Everyday Life* (Oxford: Oxford University Press, 2009), 177–79; Sturm and Nixon, "Home-Grown Social Capital," 11–12, 26, 40–41; Susan Sturm and Haran Tae, "Leading with Conviction: The Transformative Role of Formerly Incarcerated Leaders in Reducing Mass Incarceration" (New York: Center for Institutional Change, 2017), 52, https://www.justleadershipusa.org/wp-content/uploads/2017/04/Leading-with-Conviction-Report-1.pdf.

39. Daminger et al., 24–25, 35–36. On the risks of increasing inequities through technology, see Virginia Eubanks, *Automating Inequality: How High-Tech Tools Profile, Police, and Punish the Poor* (New York: St. Martin's Press, 2017).

40. See, for example, "Web Content Accessibility Guidelines," https://www.w3.org/WAI/intro/wcag; Web Accessibility Initiative, "Introduction to Web Accessibility," https://www.w3.org/WAI/fundamentals/accessibility-intro; Adobe Creative Cloud, "Making your Website Design Accessible," https://blogs.adobe.com/creativecloud/making-your-website-design-accessible.

41. See, for example, Elizabeth Cohen, "Hope for a Smoother Ride on Healthcare.gov," CNN, October 15, 2013, https://www.cnn.com/2013/10/15/health/obamacare-signup-issues-cohen/index.html.

42. Reuters, "Clinton Frees GPS Signals," *Los Angeles Times*, May 2, 2000, http://articles.latimes.com/2000/may/02/business/fi-25557; for a discussion of this and other ideas for government, see also Sunstein, *Simpler*, 80.

43. For other examples, see Anu Partanen, *The Nordic Theory of Everything: In Search of a Better Life* (New York: HarperCollins, 2016).

44. For some examples, see Emens, "Admin," 1469.

45. See OECD, *Tax Administration 2015: Comparative Information on OECD and Other Advanced and Emerging Economies* (Paris: OECD Publishing, 2015), 255–56, http://www.keepeek.com/Digital-Asset-Management/oecd/taxation/tax-administration-2015_tax_admin-2015-en#page1.

46. Ibid., 256; Deloitte, "Comparative Study of the Personal Income Tax Re-

turn Process in Belgium and 33 Other Countries," May 2015, https://www2.deloitte.com/content/dam/Deloitte/be/Documents/tax/TaxStudiesAndSurveys/TaxReturnStudy2015-EN.pdf.

47. Joseph Bankman, "Using Technology to Simplify Individual Tax Filing," *National Tax Journal* 61, no. 4 (December 2008): 783–85.

48. In a program evaluation in 2009, the Franchise Tax Board projected net savings to the state of expanding the pilot to a statewide program to be hundreds of thousands of dollars, though much of that savings was surely due to e-filing, which is by now already done for most federal returns. State of California Franchise Tax Board, "Franchise Tax Board Report to the Legislature" (April 23, 2009), 4, https://www.ftb.ca.gov/readyReturn/ReadyReturnReport2009.pdf.

49. Julie Small, "California's ReadyReturn Reduces Taxpayer Angst for a Small Number," SPCR.org, April 14, 2010, http://www.scpr.org/news/2010/04/13/13928/Readyreturn.

50. Dennis J. Ventry Jr., "Intuit's End-Run," *Los Angeles Times*, July 21, 2010, http://articles.latimes.com/2010/jul/21/opinion/la-oe-ventry-intuit-20100721.

51. See Bankman, "Using Technology," 786–88, which describes several criticisms of ReadyReturn, including potentially overstating tax liability, eliminating a financial-planning opportunity, removing an opportunity for taxpayers to participate in government, avoiding "real" tax reform by making taxpayers less angry about taxes, and exposing taxpayers to privacy and security threats.

52. See Michael J. Graetz, *One Hundred Million Unnecessary Returns: A Simple, Fair, and Competitive Tax Plan for the United States* (New Haven, CT: Yale University Press, 2008), 48–49.

53. Ventry, "Intuit's End-Run."

54. Graetz, *One Hundred Million Unnecessary Returns*, 49.

55. "CalFile Frequently Asked Questions," State of California Franchise Tax Board, https://www.ftb.ca.gov/online/calFile/faq.shtml. Under ReadyReturn, the state's Franchise Tax Board took information from forms provided by employers and banks (W-2 and 1099 forms)—which the state would have used to verify an individual's return in a self-prepared return system—and prepared a draft return for the individual. See Small, "California's ReadyReturn."

56. Ronald W. Reagan, "Address to the Nation on Tax Reform," May 28, 1985, https://www.reaganlibrary.gov/research/speeches/52885c.

57. Bankman, "Using Technology," 774; Austan Goolsbee, "The 'Simple Return': Reducing America's Tax Burden Through Return-Free Filing" (Brookings Institution, 2006), 7–10, https://www.brookings.edu/wp-content/uploads/2016/06/200607goolsbee.pdf.
58. Neil H. Buchanan, "The Easy Tax Filing System We've Been Waiting For," *Newsweek*, April 15, 2016, http://www.newsweek.com/easy-tax-filing-system-been-waiting-448357. The bill was reintroduced in 2017; see "S.912: Tax Filing Simplification Act of 2017," GovTrack, updated April 26, 2017, https://www.govtrack.us/congress/bills/115/s912.
59. See Emens, "Admin," 1464–66 (discussing the Paperwork Reduction Act, federal government attempts to reduce paperwork demands on the public, and increases in paperwork demands despite those efforts). At the local level, a law reform effort in Connecticut is trying to push lawmakers to consider the admin costs of proposed legislation and to require regulated industries to report on their "user experience." See Judiciary Committee Connecticut General Assembly, "Report of the Task Force to Improve Access to Legal Counsel in Civil Matters," December 15, 2016, 4, 23–24, http://www.rc.com/upload/O-Hanlan-Final-Report-of-CT-Leg-Task-Force-12_2016.pdf.
60. Office of Management and Budget, *Information Collection Budget of the United States Government* (Washington, DC: Office of Management and Budget, 2016), 2, https://obamawhitehouse.archives.gov/sites/default/files/omb/inforeg/icb/icb_2016.pdf.
61. Adam Samaha, "Death and Paperwork Reduction," *Duke Law Journal* 65 (2015): 295; Proposed Collection, Comment Request for Form 1040, 79 Fed. Reg. 24,500, tables 1–2 (April 30, 2014).
62. For instance, see Graetz, *One Hundred Million Unnecessary Returns*, 3–16, 197–213.
63. Abraham Joshua Heschel, *The Sabbath: Its Meaning for Modern Man* (New York: Farrar, Straus, and Giroux, 2005), 8.
64. Tiffany Shlain, interview by Krista Tippett, "Growing Up the Internet," March 31, 2016, in *On Being*, podcast, https://www.onbeing.org/programs/tiffany-shlain-growing-up-the-internet/.
65. See, for example, Maura Thomas, "Your Late-Night Emails Are Hurting Your Team," *Harvard Business Review*, March 16, 2015, https://hbr.org/2015/03/your-late-night-emails-are-hurting-your-team; Sue Shellenbarger, "A Day Without Email Is Like . . . ," *Wall Street Journal*, updated October 11, 2007, https://www.wsj.com/articles/SB119205641656255234;

Leah Fessler, "Arianna Huffington Deletes Every Email Her Employees Receive While They're on Vacation," *Quartz*, August 25, 2017, https://qz.com/1061410.

66. See, for example, "No Email Day," Plan Digital, https://plandigitaluk.com/no-email-day (declaring international no-email day); National Day of Unplugging, "National Day of Unplugging," https://www.nationaldayofunplugging.com.
67. See Holistic Life Foundation, https://hlfinc.org.
68. For some broader challenges that could help shape additional admin difficulties for kids living in poverty, and some strategies for addressing them, see, for example, Ruby Payne, "Nine Powerful Practices: Nine Strategies Help Raise the Achievement of Students Living in Poverty," *Association for Supervision and Curriculum Development* 65 (April 2008): 48–52.
69. Financial literacy is one part of proposals to bring back Home Economics, also known as Family and Consumer Sciences, which has waned as a school subject since 2002/2003 especially. Carol R. Werhan, "Family and Consumer Sciences Secondary School Programs: National Survey Shows Continued Demand for FCS Teachers," *Journal of Family and Consumer Sciences* 105, no. 4 (Fall 2013): 42.
70. For one articulation of the dream of listening education, see Avi Kluger, "Listening and Its Enemies," YouTube, May 29, 2015, https://www.youtube.com/watch?v=iR-jlgWE9GQ.
71. For discussion, see, for example, Rachel Macy Stafford, "How to Control Technology Instead of Letting It Control You," January 19, 2017, https://www.huffingtonpost.com/rachel-macy-stafford/how-to-control-technology_b_9012384.html; Kevin Kelly, interview by Krista Tippett, "The Universe Is a Question," January 18, 2018, in *On Being*, podcast, https://onbeing.org/programs/kevin-kelly-the-universe-is-a-question-jan2018/; Josephine Chu, "Arianna Huffington: Tech Should Make Us More Human, Not Less," *Thrive Global* (February 8, 2018), https://www.thriveglobal.com/stories/23136.

## Acknowledgments

1. Elizabeth F. Emens, "Admin," *Georgetown Law Journal* 103, no. 6 (2015): 1409–81.

# Index

*Index references in italics refer to text graphics.*

active waiting, 17
Adams, Rachel, 69
admaximizing/adminimizing, 151–52, 169–71
admin
  basic types overview, 8–9
  being aware of, 4–5
  dealing with unreliable or disreputable people and, 12
  as defining adulthood, xiv
  definition/description, ix, 3–7
  happy events and, 80–81
  importance of, 3–4, 99–100
  life events and, xiv, 43, 80, 191
  as the parallel shift, x, 19, 45, 100, 169
  problem examples, 5–7
  redoing, 12–13
  right time to do, 22–23
  secrets and, 5
  worst admin examples/ overview, 11–13
admin as labor
  description, 8
  significance, x–xi
  traditional chores vs., 8
Admin Avoider
  description/examples, *31*, 35–36, 164, 223
  *See also* personality type strategies/ Admin Avoider
Admin Babysitting. *See* Admin Supervising
admin cascade, 13
admin compassion, 26–27, 101–102, 197–98
admin-consulting industry (proposed), 190–91
admin coupons, 116–17
Admin Denier
  description/examples, *31*, 36–37, 154
  *See also* personality type strategies/ Admin Denier
Admin Personalities Quiz
  background, 216
  quiz, 217–21
  score interpretation, 221–22

Admin Personalities Quiz (*cont.*)
scoring instructions, 221, *221*
using, 213, 223
Admin Question, The
gift giving/receiving and, 115–18
helping others and, 114, 121–23
influencing others and, 120–21, 123–24
relationships and, 129–30, 133–35, 139, 215
*See also* goal pursuit/asking the Admin Question
Admin Savings Time, xviii–xix, 199, 205, 212
admin sharing
description, 79
examples, 70–71, 78
*See also* Admin Study Hall; Admin Supervising
Admin Study Hall (ASH)
benefits, 74–76
nonprofit organization example, 73–74
overview, 73–77, *76*
reactions to, 75–76, 77, 79
steps of, *76*
variations, 77
Admin Supervising (overview), 77–79
admin that can wreck you
causes summary, 67
divorce and, 72–73
feelings/reactions summary, 67, 68
overview, 67–81, *67*
adminimizing/admaximizing, 151–52, 169–71
"AdminOmeter," 175
*adulting*, 55
affiliative style/approach, 123–24, 211
Affordable Care Act, 186
aging-parents admin
Alzheimer's-like condition, 69–70
cartoon on, *67*, 69
description, 68–69
ethnicity and, 69
as labor, 8
*Onion, The*, and, 29
Sandwich Generation and, 68–69
Alexander, Elizabeth, 153
Allen, Woody, 97
Alzheimer's-like condition example/admin, 69–70
Amazon, 46, 94, 98, 159, 171, 178–79
American Time Use Survey (ATUS), 44–45
*Annie Hall*, 151
Any.do, 109
*Art of Listening, The* (Fromm), 143
"artist's date," 110–11
*Artist's Way, The*, 110–11
autism and insurance admin, 6
avoidance
advice on, 125
costs of, 27–28
*See also* Admin Avoider

ball-dropping. *See* strategic ball-dropping
"bandwidth tax," 19–20
Belkin, Lisa, 44
Bersani, Leo, 104
bias admin
description, 80
example, 79–80

*Black Mirror*, 53–54
bombardment admin, 101
*Bridget Jones's Diary*, 163
Buddhist teachings on gratitude, 117
Bullet Journaling, 111
bureaucracy
 advantages/disadvantages overview, 15–16
 disabilities and, 18–19
 poverty and, 26

Campbell, Tom, 189
cancellation admin, 86–87
car registration (expired) example, 5–6, 27–28, 35
Carlin, George, 96
Carnegie, Dale, 114
Castile, Philando, 27
cell-phone rules, 99, 239n6 (chap. 8)
cerebral palsy and admin, 18
Chast, Roz, *67*, 69
children/admin education (proposed), 191, 196–98, 215
children/childcare admin/kidmin
 descriptions of, 9–10, *10*, 11
 distribution/path dependence and, 57
 Flexible Spending Accounts (FSAs) and, 7
 gender and, 46, 47–48, 50, 51, 231n9, 231n13, 232n20
 helping others, 121–22
 minimizing strategy and, 170–71
 outsourcing/nannies, 32, 48, 213
 parents/millennials and, 55–56, 235n11
 reform (proposed), 184–85, 186
 relationships/gender and, 45–51, 54, 57, 139–40
 sharing and, 63
children's gifts, 115–16
Chile and tax returns, 188
chores
 admin vs., 9–11, *10*
Clinton, Bill, 186
Club Sandwich caretaking, 68
companies (proposed)
 consumer contracts/fine print and, 180–81
 wasted time of consumers, 176–78, 179
completion pleasure and admin, 106
concerted cultivation, 170
connecting/disconnecting
 email/text advice, 193, 194–95, 201–2
 Jewish Sabbath tradition and, 193–94, 201–2
 NNR (No Need to Reply), 194–95, 213
 technology-preference mismatch and, 195–96
conscientiousness and mate selection, 133–35
consumer contracts/fine print reform (proposed), 180–81
costs of admin
 overview, 17–28
 poverty and, 25–27
 quantity of admin and, 24–25
 *See also specific types*
couples and admin. *See* gender and admin; relationship/marriage and admin
coupons for admin, 116–17

Covey, Stephen, 82–84, *83*, 105, 169–70
*Curb Your Enthusiasm*, 53–54

dating/choosing a partner
  asking admin questions, 135–36
  conscientiousness and, 133–35
  life admin vs. work admin, 136–37
  strong views and, 137
  trust and, 137–38
death admin
  emotions and, 93
  example, 5
  helping others with, 119–20
  planning your own, 119–20
defaults, 56–57
*Delay, Deny, Defend* (Feinman), 181
delegation and management tasks, 45
disability
  adults/admin and, 18–19
  Alzheimer's-like condition example/admin, 69–70
  bureaucracy and, 18–19
  children/admin and, 69, 70
  dyslexia and, 70
  Sandwich Generation and, 69–71
  "social model" of, 227–28n5
  technology help and, 70
disconnecting. *See* connecting/disconnecting
discount vs. rebate, 87
distribution of admin/path dependence
  children/childcare admin/kidmin and, 57
  flexibility of admin and, 58
  information and, 58–59, 61
  problem (overview), xiii
  skills and, 61–62
  volunteering and, 60
  *See also* gender and admin; relationship/marriage and admin
divorce admin example, xiv, 72–73, 200–201
Dolmetsch, Ricardo, 6
dominant style/approach, 123
"Don't Zen on me," 100
Down syndrome and admin, 69
Dropbox, 145
Duflo, Esther, 25
dyslexia and admin, 70

education
  children learning admin (proposed), 191, 196–98, 215
  dyslexia and, 70
Eliot, George, 129
emergency-planning admin, 12
emotions as contagious, 123
empirical psychology on gratitude, 117
Ephron, Nora, 134
exercise admin, 89
extended family admin
  person doing admin/examples, 55, 59
  stickiness of admin and, 59
  *See also* childcare admin

Family Medical Leave Act/unpaid leave, 179–80
family planning services, 122–23
Fancy Hands, 161

Feinman, Jay, 181–82
financial admin
  financial-aid forms, 4
  financial crisis example, 6
  gender and, 46–47, 50
Financial Ombudsman Service, UK, 177
Flexible Spending Accounts (FSAs), 7, 183–84, 201
flower giving/receiving, 115
food workers/animals treatment, 122–23
form-lovers, 15
Fromm, Erich, 143
FSAs (Flexible Spending Accounts). *See* Flexible Spending Accounts (FSAs)
future of admin
  admin-consulting industry, 190–91
  admin education for children, 191, 196–98, 215
  admin-reducing technology, 186–88
  childcare reform, 184–85, 186
  companies wasting consumer time and, 176–78, 179
  consumer contracts/fine print and, 180–81
  health-care reform, 183–84, 186
  helping those in poverty and, 185–86
  insurance industry reform, 181–82
  Respect Our Time (ROT) rating scheme, 178–80
  tax returns/government and, 188–89
  time refunds and, 176–77
  visibility, 175–76, 189–90, 198

gender and admin
  Admin Deniers and, 37
  American Time Use Survey (ATUS) and, 44–45
  childcare/child admin/kidmin and, 46, 47–48, 50, 51, 231n9, 231n13, 232n20
  finances and, 46–47, 50
  gender-egalitarian marriages and, 45, 50
  household labor and, 48, 49, 50, 51
  maternal gatekeeping, 52
  media and, 53–54
  outsourcing and, 48–50, 141–42
  overview, 44–54
  patterns and, 54
  Pew Foundation survey and, 45–46
  power and, 52–53
  rhetoric vs. reality, 51
  same-sex couples and, 50–51
  social admin and, 46
  strategic ball-dropping and, 47, 51–52
  talking about, 51–53
  views on, 44, 45–46
  *See also* distribution of admin/path dependence; relationship/marriage and admin
gender vs. sex, 50
gift giving/receiving
  admin/advice and, 115–18, 215
  Admin Coupons, 116–17
  Admin Question and, 115–18
  transforming into not-admin, 171–72

Ginsburg, James, 63
Ginsburg, Ruth Bader, 63
Global Positioning System (GPS), 186
goal pursuit/asking the Admin Question
eliminating something else, 88, 90
experts and, 86, 90
launch date and, 87–88
losing weight and, 88–91
low admin and, 85, 90
selecting partners and, 85–86, 90
"sludge" use, 86–87
upfront admin, 85, 89, 90
goals for meetings/events, 172
going off the grid and, 21, 99–100, 183
Goleman, Daniel, 21–22
Gonzalez, Andres, 196
Google Calendar, 145
gratitude
appreciation and admin, 41–42
happiness and, 117
relationships and, 147–48, 149
thanking and, 118
gratitude impasse, 149
Greenidge, Kaitlyn, 25–26

Habiliss, 161
happiness and gratitude, 117
Harris, Dan, 98
health-care industry
autism and, 6
countries with medical savings account systems, 226n6
Flexible Spending Accounts (FSAs) and, 7, 183–84, 201
reform needed, 183–84, 186
Sweden vs. US, 187
UK vs. US, 183–84
vaccination records (children) and, 187–88
helping others
causes, 122–23
childcare, 121–22
death admin and, 119–20
gift giving/receiving and, 115–18
not burdening others, 115–16, 175–76
Herman, Ken, 44
Heschel, Abraham Joshua, 192–93
Hochschild, Arlie, x, 223n2, 227n2, 231n13, 233n27; 236n5, 239n7 (chap. 8), 242n15
*See also* second shift
Holistic Life Foundation, 196
household management and admin, 44–45, 226n8, 227n1, 230nn4–5, 232n15, 234n35

Ideas to Try list
Hacks List, 212–13
Love and Relationships, 214–15
overview, 210–15
Urgent List, 210–12
identity theft
admin and, 11, 177–78
Identity Theft Enforcement and Restitution Act (US) and, 177–78
protection companies and, 13–14
Identity Theft Enforcement and Restitution Act (US), 177–78
in vitro/IVF admin, 139
influencing others
Admin Question, The, and, 120–21, 123–24

getting better service, 123–25, 211
humanity and, 124–25
sludge and, 86–87, 120–21
information transfer
relationships and, 144–45, 214
technology and, 145
insurance industry
autism and insurance admin, 6
rationing by hassle, 181–82
reforms needed, 181–82
Intentional Study Hall (ISH)
solo ISH, 91, *92*
writing and, 91
Intuit, 189
"it's easier to just do it myself," 58–59

judgment and admin
changing to admin compassion, 101–3, 197-98
distraction/missed connections and, 21–22
invisibility of admin and, 100
not valuing admin, 101
overview, 21–24, 100–103
self-judgment and, 103
on too little admin, 97
on too much admin, 96–97
trivializing admin, 101

knowing vs. negotiating, 112–13

labor. *See* admin as labor
Lareau, Annette, 170
learned incompetence, 34, 52
*See also* strategic ball-dropping
Legal Financial Obligations (LFOs), 27–28
life admin description, ix–x, 226n8
Listening Game, 143–44, 197
LMGTFY.com, 166

managerial vs. secretarial, 21, 52, 112–13
marathon vs. sprint, 112
mate selection/admin. *See* dating/choosing a partner
"material inconvenience," 177
maternal gatekeeping, 52
matrix. *See* urgent vs. important matrix
meditation
admin and, 98–99
views of, 97–98
memory on interrupted vs. completed tasks, 20
mental energy and admin, 19–20, 87–88, 157, 142, 230n8
mental load and admin, 9, 46, 226n9
millennials and admin, 55–56, 235n1
mortgage refinancing, 4
move/utilities set-up example, 60
multitasking
admin and, 132
advantages/disadvantages overview, 20
gender and, 51
murky admin, 13–14

negotiating
knowing vs., 112–13
relationship and, 145–46
*New York Times*, 6, 46
Nigh, Asha, 6
nodes, admin and, 71, 81, 86, 211

nonprofit organization example
Admin Study Hall, 73–74
mission, 73
woman with bag of mail and, 73, 74–75
Norwood, Kimberly, 79–80
Notes app, 111–12
Nouwen, Henri, 171
Novartis, 6
nudges vs. sludge, 87–88, 90, 120, 237n6

Obama, Barack, 189
*Onion, The*, 29–30
opportunity costs, 18–19, 107
organ donations and default rule, 57
outsourcing
admin and, 160–63, 213
example/time costs, 141–42
gender and, 48–50, 141–42
nannies/admin and, 32, 48, 213
personal assistants and, 160–63

paper and admin pleasures, 14–15, 107–8, 111–12
paper debt, 150
Paperwork Reduction Act, 189
parallel shift, x, 19, 45, 100, 169
penalty-avoidance admin, 13
personal assistants, 160–63
personalities quiz. *See* Admin Personalities Quiz
personality type, 30–31, *31*, 222–23, *222*
action/feelings and, 30, *31*, 39
Admin Avoider, *31*, 35–36, 164–68
Admin Denier, *31*, 36–37, 154–57
changing after divorce, 43
changing after partner's death, 42
changing for love, 40–42
comparisons, 39
competency and, 39
as fluid, 38–39
hybrids and, 153, 223
*Onion, The*, stories and, 29–30
Reluctant Doer, *31*, 33–34, 168–72
Super Doer, 31–33, *31*, 158–64
types overview, 30–31, *31*
personality type strategies, *154*
Pew Foundation survey on gender/admin, 45–46
photo thank you, 118
Piercy, Marge, 68
Plato, 102
pleasures from admin
confessions on, 104–5
deep pleasure and, 14–15, 107
forms of (summary), 112–13
not liking vs., 105–6
overview, 106–7
paper and, 107–8, 111–12
technology and, 108–10
Pliny of Alexandria, 102
polyamorous relationships, 37, 134–35
poverty and admin
children doing, 196
costs overview, 25–27
housing and, 71–72, 236n7
legal-services clinic staff/brainstorming and, 26
pain/humiliation and, 25–26
rationing by hassle and, 185
reforms needed, 185–86
responsibility and, 25
*power hour*, 79
pre-crastination, 18, 113

process failure admin, 192, 212
prompt vs. thorough, 112

rationing by hassle, 181–82, 185
Ready Return (California), 188–89, 248n48, 248n51, 248n55
Reagan, Ronald, 189
rebate vs. discount, 87
Record of the Month Clubs, 120–21
redoing admin, 12–13
regret admin, 13
relationship/marriage and admin
"admin is marriage," 45, 134, 151
adminimizing/admaximizing, 151–52
appreciation/gratitude and, 147–48
ball-dropping/shifting and, 146–47
choosing a mate and, 133–38
contrasting examples, 131–32
discussions before marriage and, 132–33, 142
foundational moments and, 138–42
fun and, 142
gift of admin and, 147
gratitude impasse and, 149
Ideas to Try list, 214–15
information transfer and, 144–45, 214
"just tell me what you want me to do," 150–51
Listening Game/exercise, 143–44, 214
love and (overview), 130–33
"need to do"/imperial-delegation gesture and, 151
negotiating and, 145–46
outsiders making assumptions and, 60–61
outsourcing and, 141–42
patterns and, 138, 139–40, 142
planning meetings, 145, 150
pre-relationship history and, 61
relative position and, 143
sex and, 125, 134, 240–41n7
skills and, 140
starting point and, 59–60
stickiness of admin and, 62–63, 139–40
troubleshooting, 148–52
trust and, 131, 132
visibility/invisibility of admin and, 144, 214–15, 234–35n35
*See also* children/childcare admin; distribution of admin/path dependence; gender and admin; weddings
relaxation after admin, 93, 94
Reluctant Doer
description/examples, *31*, 33–34, 168
*See also* personality type strategies/Reluctant Doer
Respect Our Time (ROT) rating scheme, 178–80
retirement contributions and default rule, 57
robot and admin example, 102–3
Rubin, Gretchen, 79

Salzberg, Sharon, 183
same-sex couples admin (overview), 50–51
Sandwich Generation
aging parents and, 68–69
cartoon on, *67*, 69
description, 68
Schumer Box, 180

second shift, x, 223n2, 227n2, 231n13, 239n7 (chap. 8), 242n15. *See also* Hochschild, Arlie
secretarial vs. managerial, 112–13
*Seinfeld*, 53–54
selective incompetence, 35
  *See also* strategic ball-dropping
service requests, 123–25, 211
*7 Habits of Highly Effective People, The* (Covey), 82–84, *83*
sex
  admin doing and, 125, 134, 240–41n7
  asexuality and, 104
  Bersani on, 104
  gender vs., 50
Sheible, Lenzi, 122
Shlain, Tiffany, 193
Sisyphus admin, 13
skills
  distribution of admin/path dependence and, 61–62
  relationships and admin, 140
  stickiness of admin and, 61–63, 140
sleep loss, 18
sludge
  admin and, 86–87
  description/use examples, 86–88, , 120, 237n6
  rationing by hassle, 181–82, 185
socializing, low-admin, 170, 214
solo vs. collaborative admin pleasure, 112
sprint vs. marathon, 112
stickiness of admin
  apartment-sharing example, 56
  overview, 55–64
  relationships and, 62–63, 139–40
  skills/learning skills and, 61–63, 140
  sticky defaults and, 57
  *See also* distribution of admin/path dependence
strategic ball-dropping
  description/example, 34–35, 164
  gender and, 47, 51–52
  relationships/marriage and, 146–47
Sunstein, Cass, 87
Super Doer
  description/examples, 31–33, *31*, 157–58
  *See also* personality type strategies/Super Doer
Sweden and admin, 187
Szymborska, Wisława, 206

TaskRabbit, 161
Tax Filing Simplification Act (US), 189
tax returns
  overview, 188–89
  Ready Return (California), 188–89, 248n48, 248n51, 248n55
technology
  admin pleasures and, 14–15, 108–10
  admin-reducing technology (proposed), 186–88
  communication preference mismatch, 195–96
  disabilities and, 70, 185–86
  helping/harming, 183–84
  high tech vs. low tech, 112
  information transfer and, 145
Thaler, Richard, 87
thank-you video, 118

thanking
showing gratitude and, 117–18, 119, 176, 215
thank-you notes and, 118
thorough vs. prompt, 112
time costs
Free Application for Federal Student Aid (FAFSA) and, 4
home mortgage refinancing and, 4
insurance admin and, 6
wasting time overview, 17, 19
time refunds (proposed), 176–77
Tippett, Krista, 193
Todoist, 109–10
tracking admin
phone/apps, 108–10
pleasure and, 106–7
trans admin, 242n.8
TripTiks, 186
trivializing admin, 59, 100–101, 106
*See also* visibility/invisibility of admin
trust
Admin Avoider and, 167–68
choosing a mate and, 137–38
relationships and, 131, 132
TurboTax, 189

urgent vs. important matrix
admin and, 105
description/overview, 83–84, *83*
utilities set-up example, 60

visibility/invisibility of admin
DeVault, Marjorie, 234–235n35
future and, 175–76, 189–90, 198
gender roles and, 41, 49–50, 53, 54, 58
goals and, 85
gratitude and, 117–18, 149
helping others and, 119
as invisible labor, x
outsourcing and, 49
relationship/marriage and, xi, xii, 100, 130, 144–45, 214–15, 234–35n35
understanding/seeing, xvii, xviii, 5, 7, 23, 26, 142, 144–45, 156, 169

Warner, Jack, 166
Warren, Elizabeth, 189, 232n17
Watson, John, 102
Webby Awards, 193
weddings
advance planning benefits and, 93
last-minute planning example, 93
location and, 139
name change and, 139, 242n8
wedding-planning package vs. wedding planner, 191
Whyte, David, 114
Winnicott's "good enough" mother, 93
Wonder Woman, 3
Wunderlist, 145

yoga,89–90, 97, 238n4

Zappos, 178–79
Zeigarnik effect, 20
Zirtual, 161

© Nina Subin

Elizabeth F. Emens is the Isidor and Seville Sulzbacher Professor of Law at Columbia Law School. She earned her law degree at Yale and her Ph.D. at Cambridge. She lives in New York City.